Rainfed Farming Development in Central India

About the Authors

DR. G.P. VERMA, M.Sc. Ag. (Agra), Ph.D. (Iowa, U.S.A.) retired as Director Research services/ Director Instruction, Jawaharlal Nehru Krishi Vishwa Vidyalaya, Jabalpur, M.P. after 36 years' experience of teaching, research and extension of technology to farmers' fields. Besides merit scholarships of Agra University and U.P. Govt., he was awarded Rockefeller Foundation Fellowship for higher studies in U.S.A.

His contribution to watershed based development of Rainfed/Dry farming and reclamation and control of gullies/ravines is outstanding, and in recognition of this he bagged several assignments in India and abroad from Govt. of India, Indian Council of Agricultural Research, Consultancy organizations like Consulting Engineering Services, Water and Power Consultancy Organization (WAPCO), Indian Institute of Public Opinion and Indian Institute of Public Administration. He served as Dry Farming Expert to Govt. of Paraguay for one year (1981-82), and as a member of Team of Experts to Iraq Govt. for planning and development of dry farming. He also served as an Expert to evaluate the Intensive Extension and Research Project of Gujarat State, as well as a member of ICAR Appraisal Team for National Agricultural Research Project. He had the honour of membership of Expert Committee-cum-Jury for several awards. He has been President, Gwalior Chapter of Indian Society of Soil Science (2005-12) and Chairman, Consortium Advisory Committee, National Agriculture Innovation Project, Component III, JNKVV Jabalpur M.P. and Consultant, Rajmata Vijayaraje Scindia Krishi Vishwa Vidyalaya Project "Identification of Efficient Integrated Modules for Sustainable Management of Ravines (Chambal) and Carbon Sequestration for climate Resilience in Madhya Pradesh". He was awarded as "Agriculture Scientist-2015 by Gwalior Vikas Samiti, Gwalior and was also conferred 'Life Time Achievement Award-2015, by Jawaharlal Nehru Krishi Vishwa Vidyalaya, Jabalpur, M.P.

DR. Y.P. SINGH, PG and Ph. D. in Soil Science from Agra University, Agra, U. P. worked for 22 years in various capacities in different Institutes, such as ICAR-RCER, Patna (Bihar), Krishi Vigyan Kendra (KVK), Banasthali University, Banasthali (Rajasthan), Central Soil Salinity Research Institute, Karnal (Haryana). At present he is working as Chief Scientist and in-charge, All India Co-ordinated Reseach Project on Irrigation Water Management, Z.A.R.S., Morena, R.V.S. Krishi Vishwa Vidyalaya, Gwalior M.P. He has more than 100 research and conference papers, published in National and International Journals besides popular articles, leaflets, booklets, books, book chapters etc.

Further he is working as Co-ordinator in two projects, Principal Investigator in four projects and Co-Principal Investigator in four Projects funded by ICAR, World Bank, DST and State Government. He evolved several innovative technologies adopted by a large number of beneficiaries.

He participated in many National and International Seminars/ workshops and training programmes. He also guided six PG and Ph. D. students and got several Honours and Awards. He also worked as councilor of Indian Society of Soil Science for two years. He is also life member of five national societies.

Presently he is working as Principal Investigator in Tribal sub-plan project on management of irrigation water, National Initiative for Climate Resilient Agriculture funded by ICAR, Co-Principal Investigation of MPWSRP funded by M.P. Government and "Identification of Efficient Integrated Modules for Sustainable Management of Ravines (Chambal) and Carbon Sequestration for Climate Resilience in Madhya Pradesh" funded by ICAR.

Rainfed Farming Development in Central India

Dr. G.P. Verma
Ex. Director Research, (JNKVV, Jabalpur, MP),
108-Balwant Nagar, Gwalior-474 002, M.P.

Dr. Y.P. Singh
AICRP-IWM, ZARS-R.V.S. Krishi Vishwa Vidyalaya,
Morena-476 001 M.P.
Email- aicrpmorena@gmail.com

SCIENTIFIC
PUBLISHERS

Published by
SCIENTIFIC PUBLISHERS (INDIA)

Jodhpur –
5 A, New Pali Road
P.O. Box 91
Jodhpur - 342 001 INDIA

Delhi –
4806/24, Ansari Road
Daryaganj
New Delhi - 110 002 INDIA

ISBN: 978-93-86102-87-4

Visit the Scientific Publishers (India) website at
http://www.scientificpub.com
Printed in India
Digitally Printed at Replika Press Pvt. Ltd.

FOREWORD

It is a fact that the benefits of Green Revolution ushered in mid-sixties of the last century seem to vanish if monsoon rains are not timely and normal. Our Prime Minister, therefore, rightly emphasised the need for second Green Revolution. Fortunately seventy percent part of India receives an annual rainfall greater than 500 mm which if managed appropriately can definitely result in ushering an era of Ever Green Revolution in rainfed areas which extend over about two thirds of the total cultivated area of India, but notably suffer from twin problems of low and instable crop-yields. The prospects for the development of rainfed farming are bright as very useful information on crop-plans, their management practices has been made available by All India Coordinated Research Project on Dryland Agriculture (ICAR), land and rain-water management practices on watershed basis have been evolved by Operational Research Projects, soil and water conservation research centres, watershed development projects and State Agricultural Universities. Based on their long experience, the authors have done a commendable job of compiling the available information on land and water management, appropriate crop-plans and integrated farming systems so as to synthesize various components into a technology for the holistic and sustainable development of rainfed farming in various agro-climatic zones of Central India. Also, only those practices have been emphasized which have been successfully tested on farmers' fields.

I have every hope that the book entitled "Rainfed Farming Development in Central India" will serve as a good field manual for workers of the Department of Agriculture, as a comprehensive treatise on rainfed farming/watershed management and thus a valuable teaching material for undergraduate and post graduate courses of rainfed farming/watershed management offered by Agricultural Universities. Besides, the book gives an insight into problems and their solutions in respect of rainfed farming to those responsible for planning and guiding the development projects on rainfed farming/watershed management.

(Dr. A. K. Singh)
Vice Chancellor,
R.V.S. Krishi Vishwa Vidyalaya,
Raja Pancham Singh Marg,
Gwalior-474 002 (MP)

PREFACE

The importance of rainfed farming in Central India can not be over emphasized as development of canal irrigation had taken place at snails speed and tube-wells and open dug wells were the main source irrigation in Uttar Pradesh, Madhya Pradesh and Bihar, resulting in excessive exploitation of under ground water. Further, a significant decrease in number of tanks has aggravated the situation by a fall in recharge of underground water. This points out the urgent need of rainwater management particularly the aspect of run-off collection in different ways for tank/farm pond irrigation, domestic use, and recharging underground water. This is really the strategy of developing the rainfed farming on watershed basis. Under All India Coordinated Research Project on Dryland Agriculture launched by Indian Council of Agricultural Research throughout the country in 1969-70, good research work was conducted to identify (i) most remunerative crop/variety, (ii) nutrient requirement of crops/cropping systems, (iii) most remunerative cropping sequence, (iv) more remunerative intercropping system, (v) alternative land use and (vi) water balance studies: run-off collection and recycling. Further, the ICAR also sanctioned 23 Operational Research Projects (ORPs) to test and transfer the research results on important problems of crop production and livestock to farmers' fields. The results of All India Coordinated Research Project on crop-production and management practices, of operational research and development projects on watershed management, relevant information available in various publications, results of rain-water management research conducted by various State Agricultural Universities and other Institutes and valuable information from personal communications of many senior scientists, have all been synthesized into a technology for holistic and sustainable development of rainfed farming and presented in this book entitled Development of Rainfed Farming in Central India.

The book has 10 chapters *viz.* (1) Problems and prospects of rain-fed farming, (2) Present status and strategy of rain-fed farming development, (3) Land and water resources, (4) Technology for watershed based rain-fed farming development, (5) Land and water management practices, (6) Improvement of productivity of rain water, (7) Cropping systems and crop management practices, (8) Planning for aberrant weather conditions and drought management, (9) Farm machinery and implements and (10) Integrated farming systems for livelihood security. In support of various recommendations and conclusions drawn, reliable data have been presented in 62 tables. Further, the book is well illustrated through 24 figures/sketches.

While land and rain water management practices recommended in the book are applicable to whole country except arid regions, crop husbandry and alternate land use practices are locality specific. That is why the title of the book is "Rainfed Farming Development in Central India".

It is hoped that the book will serve as (i) a manual for field workers of the Department of Agriculture, (ii) a good teaching media for undergraduate and post graduate courses on Watershed Management/Dry farming offered by Agricultural Universities and (iii) a thought provoking material for those responsible for planning and executing development projects on rain-fed/dry farming.

G. P. Verma
Y. P. Singh

ACKNOWLEDGEMENT

We thankfully acknowledge the contributions of all the scientists who have worked or are still working under All India Coordinated Research Project on Dryland Agriculture (Indian Council of Agricultural Research) particularly those at Rajmata Vijayaraje Scindia Krishi Vishwa Vidyalaya Campus College of Agriculture, Indore, Madhya Pradesh. We are really grateful to Dr. A. Alam and Mr. V.S. Kale for their valuable contributions on 'Farm Implements' and 'Run-off management', respectively. Our heartfelt thanks are due to British Team Leaders, viz. Prof. N. W. Hudson, Soil Conservation Engineer and Prof. T. F. Shaxson, Land Husbandry Expert, who guided the scientists of Indo-British Operational Research Project (Called Indo-UK Dry Farming Project later on) to implement the land and rain water management programme of the project successfully on farmers' fields, although the required technology for Vertisol and associated soils under high rainfall conditions was not available at that time. Thanks are also due to Dr. R. P. Singh, Dr. J. S. Samra, Dr. Raj K. Gupta, Dr. Gurabachan Singh, Dr. A. K. Sikka, Dr. V. S. Tomar, Dr. V. N. Sharda, Dr. S. S. Tomar, Dr. J. C. Dagar, Dr. M. P. Jain, Engr. A. K. Singh and others whose significant contributions to rainfed farming have been cited in the book. We are indebted to Dr. A. K. Singh, Hon'ble Vice Chancellor, Rajmata Vijayaraje Scindia Krishi Vishwa Vidyalaya for his valuable suggestions and guidance after going through the manuscript and bless us with a Foreword to this book. Last but not the least, we are grateful to Mr. Alok Kulshresth, Nikhar Printing Press, Jagannath Chamber, Padav, Gwalior, M.P. for the hard work he has put in making drawings and sketches to illustrate the book through figures.

G. P. Verma
Y. P. Singh

LIST OF ABBREVIATIONS

AICRPDA	All India Coordinated Research Project on Dryland Agriculture
CGWB	Central Ground Water Board
CIMMYT	International Centre for Maize and Wheat Improvement
CRIDA	Central Research Institute of Dryland Agriculture
CSARD	Centre of Sustainable Agriculture and Rural Development
CSWCRTI	Central Soil and Water Conservation Research and Training Institute
DES	Directorate of Extension Services
DFW & AD	Department of Farmers Welfare and Agriculture Development
DFP	Dry Farming Project
IARI	Indian Agricultural Research Institute
IASRI	Indian Agricultural Statistics Research Institute
ICAR	Indian Council of Agricultural Research
ICRISAT	International Crop Research Institute for Semi-arid Tropics
IGKVV	Indira Gandhi Krishi Vishwa Vidyalaya
JNKVV	Jawaharlal Nehru Krishi Vishwa Vidyalaya
NAIP	National Agriculture Innovation Project
NATP	National Agriculture Technology Project
NBSSLUP	National Bureau of Soil Survey and Land Use Planning
NCHSE	National Centre for Human Settlement and Environment
NRAA	National Rainfed Areas Authority
RVSKVV	Rajmata Vijayaraje Scindia Krishi Vishwa Vidyalaya
RWC	Rice-Wheat Consortium
WAPCOS	Water and Power Consultancy Services
YKGPVS	Yeshwant Krishi Gram and Panlot Vikas Sansthan

LIST OF FIGURES

CONTENTS

watershed should be permissible under rainfed farming because it was appropriate use of rain water (recycled) only. No breakthrough in traditional dryfarming/rainfed farming can be achieved without appropriate management of rain-water including runoff collection and recycling.

Irrespective of definitions, the most critical input for the dryfarming as well as the rainfed farming was rain water, and the crux of the problem of development of dryfarming/rainfed farming was the appropriate management of rain-water under varying conditions of climate, land and soil. Conservation of soil moisture and collection of run-off water and recycling it for supplementary irrigation was as essential for the development of dryfarming as for rainfed farming development. Further, over vast areas in Central India, land was left fallow during rainy season, when most of the annual rainfall in these areas was received and crops were raised on profile-stored moisture in *rabi* (winter) season under moisture stress conditions in spite of high rainfall and adoption of conservation measures for natural precipitation. Therefore, no distinction could be made between dryfarming and rainfed farming on the basis of moisture stress/total annual rainfall.

1.2. IMPORTANCE OF RAINFED FARMING

Rainfed farming or dryland farming in India was important not only because it was still practiced over more than two third part of the country, but also because it supports nearly 40 percent of the human population and about 60percent of livestock population of the country. Further, it would be important for time to come because even if every drop of surface water was utilized for irrigation (not possible of course) not more than 45 percent of the cultivated land could be irrigated. The rainfed farming presently contributed 44 percent to the nation's food basket, and coarse cereals, pulses and oilseeds were almost exclusively grown as rainfed crops. Thus, rainfed farming played a major role in the agrarian economy of the country. Also, the dryland agriculture extended widely encompassing the desert terrain of Rajasthan in the northwest, the plateaus and highlands of Gujarat, Madhya Pradesh, northern Chattisgarh, Jharkhand and Maharashtra in central India, the alluvial plains of the Yamuna - Ganga river basins in the north central and rain shadow region of Maharashtra and high lands of Andhra Pradesh and Tamilnadu in Deccan Plateau (Katyal, 1992).

1.3 HISTORY OF RAINFED FARMING IN INDIA

The history of agriculture in India was the history of rainfed farming. Since the time immemorial man started taming some useful plants as rainfed farming. In the beginning, natural resources (especially land and water) were abundant and human population much less and thus whatever crop yields were obtained the produce was more than adequate to meet their requirements. With time population increased and natural resources started dwindling due to diversion of use and lack of appropriate management. Consequently the frequency and severity of extreme weather events like drought, floods, cold and heat waves and

I

PROBLEMS AND PROSPECTS OF RAINFED FARMING

1.1. RAINFED FARMING VERSUS DRYFARMING

Dryfarming and rainfed farming were often used synonymously. But a distinction between the two was made by defining dryfarming as the practice of crop production in low rainfall areas by conserving natural precipitation and application of supplemental irrigation if possible, through harvested run-off water. Rainfed farming on the other hand was defined as the production of crops in sub-humid or semi-arid regions without irrigation (Kampen et al 1976). Panda (2009) has distinguished between Dryland agriculture, Dryfarming and Rainfed farming as follows:

i. Dryfarming is cultivation of crops in areas where annual rainfall was less than 750mm and crop failures due to prolonged dry spells during crop period are most common.
ii. Dryland farming is cultivation of crops in areas with annual rainfall less than 1150mm. Dry spells might occur but crop failures are less frequent. Higher evapo-transpiration than the total precipitation was the main reason for moisture deficit in these areas.
iii. Rainfed farming is cultivation of crops in regions where annual rainfall was more than 1150mm. There was less chance of crop failures due to dry spells in humid areas where rainfed farming was practiced.

This classification apparently was based on total annual rainfall and ignored (i) distribution pattern of rainfall (number of rainy days), (ii) type of soil-texture and depth of soil and (iii) land topography which affected the moisture availability period for crop-growth very significantly and thus not appropriate. Further, the crop/variety grown by the farmer may not be suitable for his land/soil due to its long duration and can consequently fail, if prolonged dry period in rainy season occurs. As a matter of fact, both dryfarming and dryland farming are rainfed farming. Dryfarming is rainfed farming in arid, semi-arid and sub-humid regions where crops suffer from moisture stress unless rain-water was conserved through inter plot water harvesting or other means. Rainfed farming as the term suggests was cultivation of crops utilizing only rain-water and not ground-water or inter-watershed transferred water. However, life-saving or come-up irrigation through run-off collection and recycling in the same cyclones started increasing markedly in India. Frequent famines prompted the Government of India to appoint a Famine Commission in 1880 which recommended to establish protective irrigation projects for the benefit of dry and scanty rainfed tracts. It also recommended the establishment of Department of Agriculture at the national level and in various provinces (states).

The later part of 19th century saw a tremendous impetus on rainfed farming in USA and Canada. Early writers like H. W. Compbell, William Mc Donald and John A. Widtsoe inspired by the works of Jethro Tull in England advocated repeated cultivation for moisture conservation. This practice of vigorous and continuous stirring of top-soil to get a fine seedbed particularly for wheat is still common amongst the farmers practicing rainfed farming with fallow-wheat rotation. Following this single and specific system of cropping rather than diversification resulted in poor income to the farmers and no income at all in drought years due to failure of wheat crop. Another inappropriate practice followed by rainfed farmers was deep ploughing along with frequent inter cultivation which led to rapid decline of soil organic matter, deterioration of soil structure, accelerated soil erosion and ultimately resulted in poor crop yields. All these practices were followed by rainfed farmers without any scientific basis.

As already stated, Famine Commission recommended the establishment of Department of Agriculture by Central and State Governments to guide the farmers. The Commission also listed the important practices such as field bunding, summer cultivation, inter-culture, drill sowing and growing drought resistant crops like ragi, sorghum and pearlmillet. Subsequently in 1923, some research on rainfed farming was initiated.

1.3.1. Important milestones in development of rainfed farming

Various important milestones in development of rainfed farming in India are as follows:

1880 : Government of India appointed Famine Commission (after a very devastating famine in 1877).

1916 : Establishment of Department of Agriculture, Gwalior state Government, at Gwalior.

1920 : Royal Commission on Agriculture emphasized Scarcity Tract Development.

1923 : Establishment of Dryland Research Plots at Manjri.

1924 : Establishment of Institute of Plant Industry at Indore for agricultural research and development.

1933 : Research Plots established at Bijapur and Solapur.

1934 : Research Plots established at Hagari and Raichur.

1935 : Research Station established at Rohtak (Punjab, now Haryana).

1942 : Bombay Land Development Act passed.

1944 : Kanitkar, N. V. wrote a monograph entitled "Dryfarming in India."

1949-50 : Soil Conservation Department was established under Dam-odar Valley Development Corporation, Hazaribagh, Bihar.

1952-57 : Soil Conservation Research, Demonstration and Training Centres established at Dehradun, Chandigarh, Agra, Kota, Vasad, Bellary and Ootacamund (now Udhagamandalam) by Government of India, Ministry of Agriculture.

1955 : Dryfarming demonstration centres started.

1956-61 : River Valley Development projects (29) implemented in nine states.

1962 : One more Soil Conservation Research centre established at Ibrahimpatnam, Hyderabad.

1967 : All Soil Conversation Research centres transferred to Indian Council of Agricultural Research (ICAR), New Delhi.

1970 : All India Coordinated Research Project on Dryland Agriculture (24 centres) launched by ICAR.

1971 : Pilot projects for Dryfarming Development launched by Govt. of India.

1972 : Establishment of International Crop Research Institute for Semi-arid Tropics (ICRISAT) at Hyderabad.

1973-74 : First Operational Research Project (ORP) on Dryfarming Development (Indo-UK Dry Farming Project) started at Indore with British collaboration.

1974-75 : Four more ORPs started by ICAR.

1980s : Heavy emphasis was laid on watershed based development of Dryfarming.

1980-81 : Watershed Management Programme in 10 catchments of Flood Prone Rivers, launched by Govt. of India, Ministry of Agriculture.

1983 : 47 Model Watersheds selected to demonstrate watershed based dryfarming development by ICAR.

1984 : In 28 more watersheds (3.47 lakh ha) watershed based development of rainfed farming by Department of Agriculture in Karnataka, Andhra Pradesh, Madhya Pradesh and Maharashtra states through World Bank funds.

1985 : Establishment of Central Research Institute for Dryland Agriculture (CRIDA) at Hyderabad.

1986 : One more Soil Conservation Research Centre (ICAR) established at Datia, Madhya Pradesh

1987 : i. Reclamation of Chambal and Yamuna ravines (62000ha) through integrated watershed management funded by Europen Economic Commission (EEC) now European Union (EU).

ii. Drought Prone Area Programme launched by Government of India (GOI) in 615 Blocks of 15 States.

iii. Desert Development Programme by Government of India in 131 blocks.

iv. Western Ghats Development by National Planning Commission in 150 blocks.

1990s : Heavy investments on watershed development continued.

1990-91 : i. Indo-German Watershed Development Project in Maharashtra.

ii. National Watershed Development Programme (1.12 lakh ha) funded by World Bank.

iii. Integrated Wastelands Development Programme (1.12 lakh ha) funded by World Bank.

iv. Comprehensive Watershed Development Project (1.13 lakh ha) in Karnataka, Orissa and Tamilnadu funded by Danish International Development Agency (DANIDA).

v. Relmajra Watershed Project in Punjab (ICAR and Government of India).

1993 : Doon Valley Project in U.P. (1.72 lakh ha) funded by European Economic Commission (EEC) now European Union (E U).

1994 : Integrated Wasteland Development Project in 25 states by Government of India.

1995 : Indo-Swiss Participatory Watershed Development Programme (35000ha).

1996 : Comprehensive Environment Conservation Project in Kerala state (507km^2) funded by Overseas Economic Cooperation Fund (OECF), Japan.

1997-2002 : National Watershed Development Project for Rainfed Areas (NWDPRA), Integrated Watershed Development Project (IWDP), Desert Development Programme (DDP), Drought Prone Areas Programme (DPAP) etc on 72.56 lakh ha area by Government of India.

2002-2006 : Rainfed farming development through watershed development projects under Rajiv Gandhi Watershed Mission, by different States.

2006-2013 : National Agricultural Innovation Project (ICAR) Component III: Sustainable Rural Livelihood Security in 91 backward districts of India.

2006 till date : Establishment of Rainfed Area Authority by Government of India.

The most important land–mark in the history of dryland/rainfed farming research was the launch of Operational Research Projects to test and transfer the technology evolved by research to farmers' fields on watershed basis during mid–seventies at Indore, Madhya Pradesh to start with and then at Fakot (Uttarakhand), Sukhomajri (Haryana), and G.R. Halli (Karnataka). Looking to the eye opening successful results and tremendous benefits gained by the farmers themselves with guidance of research workers, large number of watershed development projects were subsequently launched funded by the

National and International agencies. The next important landmark in dryfarming/rainfed farming development was recently launched National Agricultural Innovation Project (ICAR) component III which aimed at livelihood security to dryland farmers/landless labourers through Integrated Farming Systems and subsidiary occupations.

1.4. GREAT POTENTIAL

The rainfed farming in India has a great potential due to its richness in natural resources important for agriculture *viz.* land, water, flora and fauna. The country has 143 million ha of productive land and 70 percent of it receives more than 500 mm rainfall annually. Thus, only 30 percent part of the country is arid receiving less than 500 mm rainfall and the remaining 70 percent was semi-arid, sub-humid and humid receiving 500 to more than 1500 mm annual rainfall. Besides 400million ha m of rain-water, the country has vast amount of surface water (115 million ha m) stored in reservoirs, and flowing in rivers, rivulets and live *nalas*, and underground water (45 million ha m). The country enjoys a widely varying climate suitable for a large variety of crops and a huge reservoir of natural vegetation providing an enormous variation in germplasm - a boon for Plant Breeders. If these natural resources, particularly the rain-water which was the sole source of water to rainfed crops were managed properly, it would at least double the crop yields, stabilise them at a high level, minimize or eliminate the bad effects of drought and floods, and bring as much prosperity to rainfed farmers as enjoyed by those having irrigated land, at least on 70 percent part of the country (Katyal, 1995). Besides rain-water management, suitable crops, cropping systems and farming systems for rainfed farming in arid, semi-arid, sub-humid and humid regions have been found out by dryland researchers. Thus, complete technology to develop rainfed farming was now available, and the great potential of rainfed farming could be exploited on sustainable basis and an ever green revolution ushered in the country (Verma, 1980b).

1.5. PROBLEMS OF RAINFED FARMING

1.5.1. Uncertainty of rainfall

The rainfed farming/dryfarming in India, suffered from twin problems of low and instable productivity, although 70 percent part of the country received more than 500 mm annual average rainfall. The rainfall was highly seasonal and about 75 percent of the total rainfall occurred during June to September and July and August were the months of highest rainfall generally. Also, the rainfall had a very high coefficient of variation. It varied considerably in time and space with regard to its quantum, onset and recession, and occurrence of dry spells between monsoon rains were highly unpredictable and common even in high rainfall areas resulting in moisture stress leading to low and uneconomic crop-yields. The frequency of drought was once in 3 to 6 years in different parts of the country. The history of dry farming in India was that of droughts, famines and floods unfortunately.

1.5.2. Variation in rainfall intensity

The effectiveness of a particular rainfall event depends upon the rainfall intensity. Rainfall of low intensity for longer period was more effective than intense rains for a short period, in respect of (i) moisture availability to plants, (ii) damage to soil by run-off, and (iii) amount of profile stored moisture. High intensity rain results in more run-off and soil erosion, particularly on bare and slopy land. Some high intensity rains occur in low rainfall areas also.

1.5.3. Inappropriate crop plans

The drought was not due to inadequate rains alone but also because of inappropriate crop-plans adopted by farmers, irrespective of moisture availability period in different agro-climatic regions. To be precise, at many places, the rice crop grown on sandy and shallow soils and on slopy lands was responsible for frequent droughts.

1.5.4. Low soil fertility status

Besides availability of moisture, the misery was aggravated by impoverishment of soil by water-erosion caused by high intensity rains in some rainfed areas particularly where the land was left fallow in *kharif* season. Degree of severity of water-erosion was directly related to decrease in crop-productivity. Sehgal and Abrol (1994) reported a decrease of 5 to 50 percent due to water erosion of major soil groups of India. Also loss of productivity was more pronounced in red soils followed by black and alluvial soils. Vittal et al (1990) reported that in Alfisols, yield losses were 138, 84 and 51 kg ha^{-1} in case of sorghum, pearlmillet and clusterbean, respectively for each cm of top soil removed by erosion. Yadav et al (1993) studied the loss in crop-productivity vis-a-vis different degrees of water erosion and found that crop yields were reduced by about 10 per cent when the soil loss was slight, by 10 – 35 percent when it was moderate and 60 – 75 percent when the same was severe. Also, long duration sorghum suffered the most followed by soybean and pigeonpea while cotton was affected the least.

1.5.5. Fertilizer application not common

The soils in rainfed areas are not only thirsty but hungry too and need regular application of fertilizer and manure. But farmers hesitated to use fertilizer to *kharif* crops as they thought growing *kharif* crops was risky and *rabi* crops did not respond highly to fertilizer application due to limited profile stored moisture available for their growth. Further there was no awakening amongst the farmers about balanced fertilization/integrated nutrient management and some farmers still used urea alone. It is the imbalanced use of fertilizer by the farmers that has brought a bad name to the fertilizer that fertilizer-use harmed the soil.

1.5.6. Management problems of Vertisols and associated soils

Vertisols and associated soils posed serious problems of management due to their characteristics such as formation of deep and wide shrinkage cracks on drying and swelling on wetting, low water-infiltration rate when saturated and consequent water-stagnation and waterlogging. Waterlogging reduced the yield of *kharif* crops or even led to their failures unless appropriate remedial measures were taken. Further, such soils could be worked upon within a very narrow range of moisture content, as these got too hard to operate when dry and too wet to handle if moisture percentage increased beyond a certain limit. This necessitated operations within a limited and short time and thus a high power requirement of the farmers.

1.5.7. Other limitations

Besides constraints of deficit and excess of moisture during *kharif* season, crop production in rainfed farming suffered from other limitations such as non-adoption of high yielding short duration varieties of crops, responsive to improved package of practices, resistant to pests and diseases, and tailored to moisture availability periods in various agro-climatic regions. Also, there was no awakening amongst the farmers about integrated nutrient management (INM) and integrated pest management (IPM).

1.5.8. Lack of livelihood security

It was necessary to adopt mixed farming/integrated farming system by rainfed/dryland farmers so that they had a second string to their income. Keeping of dairy animals was quite common and appropriate too for them and therefore, food, fodder and fuel became their basic needs. While growing crops, food was priority number one, production of fodder came only next. Due to scarcity of fodder in dryland areas, the condition of livestock was pitiable and farmers had given up keeping a pair of bullocks. Deprived of reliable source of power with them, they were unable to carry farm operations timely and precisely which was so crucial in rainfed farming. Shortage of fuel wood was yet another constraint. The farmers were compelled to make cow dung cakes for fuel and thus burn the valuable organic manure so essential for their fields, particularly under rainfed farming.

1.5.9. Failure of cooperation movement

Failure of cooperative societies had proved to be a big stumbling block in the adoption and diffusion of technology evolved by research. Non-availability of inputs (good quality seed of right variety and fertilizer, etc.) and credit at reasonable rates, lack of crop insurance, prevalence of small and fragmented land holdings, unfavorable land tenure policies, poor marketing and infra-structural facilities and inadequate price support have all added further to disincentives to adopt improved dryland/rainfed farming technology and set into motion a vicious circle-low productivity resulting in low and uncertain returns and a poor resource base of dryland/rainfed farmers.

Because of these innumerable problems, rainfed farming fell a prey to drought, famines and floods inflicting misery to human beings and livestock. Accordingly since long, continued efforts had been made to devise technology for improving dryland/rainfed farming.

1.6. EARLY RESEARCH EFFORTS

It was fortunate that the misery inflicted by droughts and famines caught the attention of early rulers and the succeeding governments. Charities and gifts of food were given away to public by early rulers. Free distribution of food grains, fodder and water in some areas in some years and waiving of land-revenue were still adopted as immediate relief measures by the Government to mitigate or reduce the adverse effects of droughts. Some of the Princely States contributed to agricultural research by establishing Agricultural Research Laboratory, Research Farms and Research Institutes. As early as 1916, the Department of Agriculture and the Agricultural Research Laboratory were established at Gwalior by Gwalior State Government. Dr. Sam Higginbottom (founder of the Agricultural Institute, Allahabad) took over as the Director of Agriculture with Mr. Don W. Griffin as Agricultural Engineer, Mr. L.A. Kenoyer as Economic Botanist and Dr. R.L. Pendleton as the Chief Chemist of the Agricultural Research Laboratory. Research work for selection of varieties, method of cultivation, effect of addition of farm yard manure and tillage was conducted on important rainfed crops viz. sorghum, cotton, pearlmillet, groundnut, pigeonpea, sesame, wheat, chickpea and linseed.

In 1924 Institute of Plant Industry was established for Agricultural Development of all the princely states of Central India (now south west Madhya Pradesh) and Rajputana (now Rajasthan). The Institute conducted pioneer work on cotton, method of composting and farm implements. Indore method of composting and Indore ridger for ridging groundnut and for intercultural operations in kharif crops are the outstanding achievements of the Institute. Sir Albert Howard, Dr. J.B. Hutchinson, Dr. P.A. Gadkari, Dr. V.G. Panse, Dr. Yashwant Wad and Dr. K. Ramiah were the distinguished scientists associated with the Institute. The research work conducted in Madhya Pradesh during pre-independence period (1916-47) has been compiled by Rajmata Vijayaraje Scindia Krishi Vishwa Vidyalaya (RVSKVV, 2012). However, no effort was made to conduct systematic research on dryland agriculture to evolve long term measures for drought proofing and stabilizing the agricultural production in different agro-climatic regions of the country.

It was only in 1923 that some long term measures were thought of to improve dryfarming so as to prevent famines and reduce effects of droughts. The first research effort towards this was made by setting up small research plots at Manjari near Poona (now Pune). This was followed by a more comprehensive research scheme on dry farming sanctioned by the Imperial (now

Indian) Council of Agricultural Research during 1933-35, for low rainfall areas, *viz.* Rohtak in Punjab (now in Haryana), Solapur and Bijapur in Bombay state (now in Maharashtra and Karnatak states, respectively), Raichur in Hyderabad state (now Andhra Pradesh) and Hagari in Madras state (now Tamilnadu). Research work under this scheme at 5 centres was conducted for about 10 years (1933-34 to 1943-44), and it resulted in recommendations of (1) Bombay method of Dry farming, (2) Hyderabad method and (3) Madras Dry farming practices. The common features of these recommended practices were as follows:

1. Contour bunding of fields for prevention of soil erosion.
2. Deep ploughing once in three years for better infiltration of water into the soil.
3. Use of farm yard manure to supply plant nutrients once in three years.
4. Use of low seed rate and wider row spacing.
5. Inter-culture in standing crops for increasing the efficiency of limited moisture stored in soil, and
6. Adoption of mixed cropping and crop-rotations.

These were all generalized recommendations aiming mainly at soil moisture conservation irrespective of soil types which varied significantly at different research stations. Also, the recommended soil and moisture conservation practices could increase the yields by only 15 to 20 percent over the base yield of 200-400 kg ha^{-1}, as these recommendations were not integrated with more remunerative crop plans. Further, there was lack of plant material responsive to conserved soil-moisture and capable of stabilising the crop yield. Therefore, farmers were not enthusiastic to adopt these recommended practices.

1.7. ESTABLISHMENT OF SOIL CONSERVATION RESEARCH CENTRES

Another landmark in the history of dryland research was the establishment of nine soil conservation research centres in different parts of the country in 1954 to improve the productivity of drylands (IARI, 1976). Large scale action programmes in soil and water conservation were taken up which benefitted the dry farming areas by conserving soil and water. But a breakthrough in productivity of dryland areas was not achieved as soil and water conservation practices were not integrated with more remunerative crop-plans, and cultivation of traditional crops and their long duration varieties continued without getting full advantage of the conserved water and soil. Since available plant material could not respond to improved management, the yield increase was only marginal. Thus, soil and water conservation research did not contribute to high and stable yields of dryland crops. In mid-sixties, the photo-insensitive, short duration and high yielding varieties ushered an era of scientific agriculture in irrigated areas. Thus, while Green Revolution took place in irrigated areas, the rainfed/dryland areas were deprived of it, because soil conservation research centres and dryland research centres worked in isolation and an integrated watershed development technology for improving dryland agriculture could not be evolved.

1.8. INITIATION OF ALL INDIA COORDINATED RESEARCH PROJECT ON DRYLAND AGRICULTURE AND ITS DEVELOPMENT PHASES

As already stated, high yielding varieties of wheat and rice were introduced in mid-sixties. Fortunately, short duration and high yielding hybrids of sorghum and pearlmillet were also evolved. These hybrids performed as well under rainfed/dryland farming as on irrigated lands, not only in good rainfall years but also in drought years with appropriate management of rain water. Obviously, the need for a multidisciplinary comprehensive research project was felt strongly to deal with low productivity and instability of crop-yields on drylands. Accordingly, Indian Council of Agricultural Research sanctioned an All India Coordinated Research Project on Dryland Agriculture in 4th Five year plan (in June 1970) with an outlay of Rs 189 lakhs enhanced by a support (16 lakhs Canadian $) from the Govt. of Canada. The project had 24 centres to start with but now 25 (19 main centres, 3 sub-centres and 3 voluntary centres) throughout the country. Of these, 8 centres are located in Central India. Two main centres are located in Madhya Pradesh at Indore and Rewa. Four centres are located in Uttar Pradesh-one main centre at Varanasi, two sub centres at Agra and Faizabad and one voluntary centre at Jhansi. One main centre each in Chhattisgarh and Jharkhand are located at Jagadalpur and Ranchi, respectively. Simultaneously twenty four Integrated Pilot Development Projects for Dryland Agriculture were initiated by Govt. of India, with an outlay of Rs. 200 million. Both dryland research project and the pilot development project were strengthened and continued under 5th five year plan with outlays of Rs. 300 lakhs and Rs. 1000 lakhs, respectively. All the 25 research centres are still continuing (Fig. 1).

1.8.1. First phase of Dryland Research

In the first phase of dryland research uniform type experiments were conducted at all centres to determine (i) the most remunerative crop/variety, (ii) efficiency of different crops in utilizing rain-water/profile stored moisture, (iii) most suitable crop-sequence and crop mixture, (iv) optimum crop geometry for most remunerative crop/variety, (v) water intake rate of soil, (vi) water balance studies using lysimeters, (vii) most economic and practicable mulch, and (viii) fertilizer use in drylands. Experiments were also conducted to evaluate pre-release materials and to evolve more suitable varieties of the most efficient crop. Run-off and its recycling studies were also conducted at some centres. Lot of valuable information was collected during initial three years and it was established that yields of dryland crops could be increased by 100 percent or more by selecting appropriate variety, sowing it in time, controlling weeds, managing rain-water appropriately and using fertilizer. Cropping intensity could be increased by intercropping and by sequential cropping in better rainfall areas or in better rain-distribution years.

1.8.2. Second Phase (1974-80): Initiation of Operational Research Projects (ORPs)

Transfer of technology to farmers fields has been a matter of concern to ICAR from the very inception of dryland research. Very good results were obtained with regard to more remunerative crops, their varieties, improved package of practices, crop geometry, tailoring crop-plans to moisture availability periods in different agro-climatic regions, tillage and seeding practices and alternate land-use and run-off collection and recycling for high and stable production. But the transfer of this technology to farmers' fields through Pilot Development Projects executed by various state Departments of Agriculture was rather slow, and not proper in some cases, e.g. field bunding and property bunding was undertaken in the name of contour bunding irrespective of soil type. Therefore, Indian Council of Agricultural Research sanctioned four ORPs to transfer and test the technology evolved by dryland research workers to farmers' fields and to study the operational problems in the transfer of technology by the scientists themselves. The first ORP on dryland agriculture (Indo-UK Dry Farming Project) was initiated at Indore (Madhya Pradesh) in 1974 with an outlay of Rs.40 lakhs from ICAR and an assistance of about £ 5000 for equipment etc. from the U.K. Government. Subsequently, three more ORPs sponsored by Canadian International Development Agency (CIDA) were started at Sukho-majari (Haryana), Fakot (Uttarakhand) and GR Halli (Karnataka). The ORPs did an eye-opener job of technology transfer for the development of dryfarming on water-shed basis.

1.8.3. Third phase of Dryland Research

The third phase (1980-85) was an important milestone in dryland research due to change in pattern of experiments and intensification of technology transfer effort through establishment of Operational Research Projects. Throughout the country in this phase, emphasis was shifted from type experiments to location specific projects and systems approach was focused. Efforts were directed towards full exploitation of agro-climatic resources, manipulation of sowing and harvesting dates, identification of genotypes and cropping systems that could ensure stable optimal yields and maximize profits. Efficient management of inputs like fertilizers and manures continued to receive attention. Precision seeding and placement of fertilizers in relation to moisture in the seeding zone was emphasized and achieved through bullock drawn seed-cum-fertilizer drills through attachments developed for the purpose. Year-round tillage as a means to control weeds was conceived as very important step. Strategies for aberrant weather conditions were refined. Studies on alternate land uses and crop plans according to suitability of land, pasture development, fodder crop, etc. were undertaken (Singh, 1987).

1.8.4. Fourth phase of Dryland Research (1985-2000)

In view of the commendable research work done at various centres of All India Coordinated Research Project on Dryland Agriculture (AICRPDA), and looking to the urgency of perfecting the technology for dry farming development – consisting of land husbandry, crop husbandry and animal husbandry practices, the Indian Council of Agricultural Research elevated the status of directorate of AICRPDA to full-fledged Central Research Institute of Dryland Agriculture (CRIDA) in April, 1985. The Institute has four divisions, viz. (i) Division of Crop Science, (ii) Division of Resource Management, (iii) Division of Design and Analysis, and (iv) Division of Transfer of Technology. Meanwhile the Operational Research Projects on dryfarming started in 1974 had already adopted the appropriate approach for developing the dryfarming on farmers' fields on watershed basis and pointed out the need for locality specific basic information on land, soil, climate, hydrology, water-balance, moisture availability period, etc. This further strengthened the case of establishment of research institute for dryland agriculture. Various centres of AICRPDA planned all their experiments in an independent watershed, so that an integrated watershed development technology was evolved after all the required basic information was collected through appropriate studies. Very fruitful research results specific to different agro-climatic regions have been obtained by ACRPDA on characterization of environment, evolving dryfarming technology matching the environment, contingent crop planning to reduce the adverse effects of aberrant weather in drought prone areas, making sustainable utilization of land and water resource to bridge the gap of prosperity between drylands and irrigated areas, particularly where average annual rainfall was more than 500 mm.

1.8.5. Fifth phase (2000 onwards)

In the current phase, studies on characterization of environment and contingent crop- planning for different degrees of drought and for drought at different stages of crop-growth have been undertaken. Based on local conditions of soil and rainfall, suitable crop sequences and cropping systems tailored to moisture availability period were worked out. The emphasis laid earlier on run-off collection and its recycling has been accelerated through studies on rain water management, its conservation, most efficient use, catchment - storage – command relationship and micro-irrigation methods. Based on feed–back from operational research projects, studies were intensified on alternate land–use i.e. agri-horticulture systems, silvi-pastoral system and inclusion of fodder and forage crops in remunerative cropping system to enhance quality and quantity of fodder for livestock. Experiments on tillage, weed-management, mulches, use of residual soil moisture after *kharif* cropping, water-use pattern of different crops/varieties and long term use of plant nutrients and integrated nutrient management for rainfed crops were being conducted at different research centres.

On the front of extension of dryland technology to farmers' fields, huge investments were made on development of rainfed farming on watershed basis through numerous nationally and internationally funded projects. Recently concerted efforts have been made under National Agricultural Innovation Project (2006-13) component – III implemented in 91 backward districts of India to develop rainfed farming through integrated farming systems to assure sustainable rural livelihood security. Further, the Government of India has very recently commissioned the National Rainfed Area Authority (NRAA) under the planning commission to lay heavy emphasis on development of rainfed farming. The NRAA has assigned the task of prioritization of rainfed area to Central Research Institute for Dryland Agriculture (CRIDA), Hyderabad and Indian Agricultural Statistics Research Institute (IASRI) New Delhi. The joint team of scientists from both these Institutes has come up with Rainfed Areas Prioritization Index (RAPI). Based on RAPI score, the study identified 167 districts as the top one third among 499 as prioritized districts. Again, of these 167 prioritized districts, 50 need immediate attention for enhancing productivity and livelihood as resource-wise they are rich but the productivity and livelihood status are poor (CRIDA and IASRI, 2012).

II

PRESENT STATUS AND STRATEGY

India receives rains through monsoon winds from both Arabian Sea and Bay of Bengal, but the distribution of rain is not uniform over time and in space. In central India too, the rainfall varies considerably not only in different parts but also from year to year. While in some years the rainfall received in the states was almost normal, in other years the same was either above average or abnormally low. Similarly, while in some parts the rainfall was above average, in other parts the same was quite low resulting in drought in these areas of the states. On the whole, the annual rainfall was not only uncertain but highly variable and erratic in distribution. Since rainfall was the only source of water to crop under rainfed conditions, the rainfed farming was more or less a gamble with the monsoon.

2.1 LAND AND RAIN WATER IN CENTRAL INDIA

Central India consists of Madhya Pradesh, Chhattisgarh, Uttar Pradesh, Bihar and Jharkhand states (Fig. 1). The geographical area of Central India was 85.82 million ha, of which 48.96 million ha (57.0 percent) was cultivated as against national figure of 43.5 percent. The vast cultivated land in the region was blessed with highly productive soils of different types, *viz.* Inceptisol, Entisol, Alfisol, Mollisol and Vertisol and associated soils (medium and shallow black soils).

The average annual rainfall in central states varied from less than 700 mm in the west to 1600 mm in the east *i.e.* it increased from west to east. In spite of the adequate annual average rainfall, the agricultural production under rainfed conditions suffered from twin problems of low productivity and instability due to low to scanty and untimely rains in some part, and erratic distribution in other parts. There was ample underground water also, though varying in amount in space. Thus the states had very productive land and adequate water resources, but unfortunately the utilization of land and water resources in the states was very poor and in-appropriate as was quite apparent from the existing

i. Land and water (rainfall) management practices,

ii. Cropping pattern and agronomic practices, and consequently

iii. Low yields of rainfed crops obtained in the central states.

Fig. 1. Central India and Dryland Research Centres

2.2 PRESENT LAND AND RAIN-WATER MANAGEMENT PRACTICES

Mainly two systems of cultivation were being adopted in the states by farmers *viz.* upland cultivation and low land cultivation, which were also called open field and haveli/bandhi system of cultivation, respectively.

2.2.1 Open field/upland cultivation

This system of cultivation depended mainly on the type of soil and system of external drainage. Generally the land with light and medium soils and gradual slope has good drainage whereas flat land with heavier soils has poor drainage. Under good drainage conditions, fields were prepared and crops sown either adopting dry sowing during the pre-monsoon period or as post-monsoon sowing after receiving adequate rain. Intercultural operations were given for controlling the weeds as and when needed and permitted by soil condition. The heavier soils with good drainage can be very well used for raising *kharif* crops. Wherever drainage was not adequate, no *kharif* crop except paddy could be grown under rainfed conditions. Consequently, vast areas with heavy deep black soils were left fallow in *kharif* (monsoon season) and frequent cultivations given to conserve moisture and control weeds throughout the monsoon season on attainment of the appropriate condition of the soil moisture (called *batar* locally) to permit tillage operations. Thus only one crop either in *kharif* or in *rabi* was grown, following either two-year crop rotation: *fallow-rabi, kharif-fallow*, or one year rotation: *fallow-rabi.* When the land was left *fallow* during monsoon, it was subject to severe erosion particularly when the slope was more than one per cent.

2.2.2 Bunded field cultivation

Bunding has been practiced quite extensively in Madhya Pradesh as a soil and water conservation measure. Depending upon the topography and soil condition, three different systems of cultivation are presently followed by the farmer under bunded conditions, *viz.* (a) haveli system, (b) bandhi system, and (c) bunded rice fields.

2.2.2.1 Haveli system: In haveli system of cultivation, water was impounded in bunded fields in rainy season to obliterate weeds. The height of bunds varied from 30cm to more than one meter. When sowing time of wheat approached, the field was drained out and wheat was straightaway drilled without any preparatory tillage. In this way, the farmer took only one crop a year in *rabi* on profile stored moisture, which varied with texture and depth of the soil. Haveli system existed on comparatively level land (slope less than one percent) with deep black soil (Vertisol).

2.2.2.2 Bandhi system: This system was almost like the haveli system but was practiced in areas with slopy land and high rainfall where severe rill and gully erosion was observed under unbunded conditions. In order to prevent the rill and gully erosion, big bandhis (bunds) across the slope were constructed so as to impound water against bandhis and check erosion. Under such conditions, upland crops (maize, sorghum, *etc.*) were grown on the upper reaches of the catchment, and rice crop below the bunds, whereas a *rabi* crop was grown on the submerged area after the impounded water was drained away and the soil came in condition. This, however, led to frequent delays in sowing *rabi* crops particularly in lower reaches due to probable flow of water from upper bandhis to the lower bandhis. This was because the drained water from upper bandhis ran off to lower ones. Apparently, the farmers on the lower reaches of a slopy watershed had to wait for sowing of *rabi* crop till all the bandhis above his holding/bandhi had been drained out. This called for an independent and adequate drainage system from each bandhi, to improve the bandhi system.

2.2.2.3 Rice fields: In many areas, slopy and undulating land had been flattened and rectangular bunded fields constructed by using bulldozer at a great cost. While it was alright in high rainfall areas and on land with heavy soils, it was not appropriate on light soils and in areas with low and uncertain rainfall as for such areas, rice was not an appropriate crop to be grown.

Table 1. Area cropped in *kharif* in some districts of Madhya Pradesh

S.No.	District	Area cropped in *kharif*			
		1976-77		2010-11	
		,000 ha	% of total cultivated area	,000 ha	% of total cultivated area
1	Raisen	64.0	15.9	120.7	47.8
2	Vidisha	115.5	22.8	180.7	64. 5
3	Sagar	143.0	27.6	259.2	70.5
4	Datia	38.5	29.0	58.0	34.2
5	Bhind	113.7	32.1	64. 5	17.9
6	Narsinghpur	92.7	34.1	123.9	78.3
7	Hoshangabad	152.9	34.5	227. 6	93.9
8	Gwalior	92.6	34.7	66. 5	22.4
9	Bhopal	55.2	36.2	98.8	89.5
10	Damoh	196.8	36.5	143.8	87.8
11	Morena	106.8	39.9	92.1	39.2
12	Satna	168.3	42.8	167.6	88.8
13	Sehore	161.0	44.1	301.0	85.9
14	Panna	100.2	44.1	86.3	61.2
15	Jabalpur	226.3	44.2	126.9	89.4
16	Guna	284.7	49.5	196.0	90.2
17	Rewa	64.0	51.9	210.4	94.3

Source: M.P. Agricultural Statistics 1977 and 2011, Directorate of Agriculture, M.P. Govt. Bhopal (M.P.)

2.2.3 Run-off water not recycled

Under all the above mentioned bunded field cultivation systems, the run-off water though collected against bandhis/bunds was not recycled for irrigation and thus these systems lacked an efficient use of rain-water. In many regions well drained fields quite suitable for upland crops have been converted into rice fields by increasing height of the bunds on all the four sides of the field. As a matter of fact, soil conservation and field-bunding had been synonymous in the states. On large tracts with heavy soils, bunding led to waterlogging and water-stagnation, and unless adequate surface drainage was provided, it resulted into either failure or considerably reduced yield of *kharif* upland crops, e.g. maize, sorghum and soybean.

2.2.4 *Kharif* fallowing

A vast cultivated area was fallow in *kharif* even under open field cultivation system and only one crop was grown in *rabi* on profile stored moisture (Table 1).

The practice of *kharif* fallowing was very serious before spread of soybean crop particularly in Madhya Pradesh. Not restricted to any particular agro-climatic zone, it was common to many zones, e.g. plateaus of Malwa, Vindhya, Kymore and Satpura, and to Central Narmada Valley and Gird region of Madhya Pradesh state.

2.2.5. Soil Erosion

Severe erosion by intense rains during the monsoon season and the consequent run-off water particularly when there was a loose layer of soil on the surface after summer cultivation and an almost saturated and impervious soil-layer under the surface soil. The rills and washes through the fields resulted in the formation of gullies, as appropriate measures were not adopted to stabilise the washes. Thus not only top soil but the valuable land also was lost every year, at a faster rate with time, as gullies deepened and widened due to bed erosion at first and then falling away of side-banks of the gully. Further, these gullies advanced into the cultivated fields with time. The extent of gully formation depended on erodibility of soil and erosivity of rainfall. On highly erodible soils having a high dispersion coefficient, a net work of gully, sub gully and tertiary gullies was formed and the whole land was changed into ravines, for example, on alluvial soils of Chambal river valley (Verma, 2000).

To sum up, very poor management of land and water resources was adopted at present on farmers' fields resulting in gross wastage of not only rain-water but also of valuable land.

2.3. PRESENT CROPPING PATTERN AND PRACTICES

The agriculture in major part of the states was rainfed, as only some parts had the assured irrigation. With varied conditions of soil and climate, the states presented very favourable conditions for a wide variety of crops. As already mentioned, presently the states had six major crop zones, viz. (i) rice zone, (ii) rice-wheat zone, (iii) wheat zone, (iv) sorghum-wheat/mustard zone, (v) maize-wheat zone, and (vi) sorghum-cotton zone. Thus, the biggest emphasis had been on rice, wheat, sorghum and cotton crops in the states, and it appears that the sole basis for the traditional choice of these crops was the total annual rainfall received in various parts of the states. For example, the rice crop was concentrated in areas receiving a total annual rainfall of 125 cm or higher. The rice-wheat zone, a transition zone between the rice zone and the wheat zone covered the area with a total rainfall greater than 100 cm but less than 125 cm. In this zone, rice and wheat crops were grown in alternate years, mostly in bunded fields. Wheat zone was confined to central flat lands with deep black soils. Sorghum-wheat/mustard zone was situated in the northern and north-western area receiving less than 100 cm total rainfall, whereas maize-wheat zone was situated in the same rainfall region in the southern part of the states. The sorghum-cotton zone was confined to comparatively lower rainfall regions of western Madhya Pradesh (600-800 mm average annual rainfall). Unfortunately, cotton was being further replaced by wheat crop in this zone i.e. the land was left fallow in *kharif* and wheat crop grown in *rabi* on residual moisture, and this was the least efficient use of land and water resources. However, there was a desirable change recently (brought about by adoption of rainfed farming technology recommended by research) with the introduction of soybean crop which was spreading considerably fast, and was reducing the *kharif* fallow-an undesirable practice. It was desirable to spread soybean further on black soil.

Table 2. Area and productivity of food grain crops

	Area (M ha)		Productivity	
State	**2000-01**	**2006-07**	**2000-01**	**2006-07**
U.P.	20.30	20.04	2105	2057
Bihar	7.12	6.70	1694	1656
Jharkhand	1.84	2.38	1095	1550
M.P.	10.78	11.78	945	1167
Chhattisgarh	4.93	5.06	589	1148

2.4. LOW YIELDS OF RAINFED CROPS

In spite of fertile land and adequate water resources the yields of rainfed crops in the central States were not only quite low but also fluctuated tremendously from year to year. The rainfed farming in central States was dominated by cultivation of food grain crops. The area and the productivity of these crops during 2000-01 and 2006-07 are shown in Table 2. The data revealed that there was a slight decline in hectarage as well as productivity of food grain crops in Uttar Pradesh and Bihar but an increase in both hectarage and productivity in Jharkhand, Madhya Pradesh and Chhattisgarh states. The increase in productivity was quite significant in Jharkhand (41.5 percent) and Chhattisgarh (94.9 percent). However, the productivity of food grains in 2006-07, was the highest in Uttar Pradesh and the lowest in Chhattisgarh state. Compared to the national productivity of food grains (1697 kg ha^{-1}), only that of Uttar Pradesh was higher where as those of other states were lower. The average yields of all crops except sorghum in Madhya Pradesh were lower than the national averages while the yield of groundnut and of chickpea was about the same as national average (Table 3). The yields of *kharif* crops were low due to (i) inappropriate or long duration of crops/varieties like cotton and desi and improved varieties of sorghum, (ii) generally delayed planting of *kharif* crops, (iii) non-application of any fertilizer, (iv) water-logging and water-stagnation in the initial stages, particularly under bunded conditions and on the heavy soils, and (v) the moisture stress during crop growth as a result of dry spells during the rainy season or early cessation of monsoon. The yields of *rabi* crops were also poor because of:

i. Limited quantity of profile stored moisture,
ii. Poor fertility status of eroded soil,
iii. No fertilizer/manure application,

iv. Late sowing of crops due to non-availability of appropriate condition of the land (water stagnation in fields for longer time), and

v. Non-adoption of improved and disease resistant varieties of wheat, chickpea, linseed, etc.

Thus, in spite of productive land and adequate water resources, farmers of central India, particularly those of Madhya Pradesh and Chhattisgarh states get only one poor crop a year under rainfed conditions. Monsoon crops suffer either from excess of rain water (water-logging and water-stagnation) or from drought due to wide dry spells, while *rabi* crops invariably suffer from inadequate moisture supply. Consequently, the farmers of the state are poor and exposed to risks of uncertainty and erratic nature of monsoon rains both in amount and distribution. They are struggling with the nature to practice monoculture of wheat, pulses and oilseeds on residual moisture in *rabi* season, on large areas.

Table 3. Comparative yield (kg ha^{-1}) of important crops

Crop	Yield in Madhya Pradesh		All India average		The highest Yield in the country		State obtaining the highest yield	
	1976-77	2006-07	1976-77	2006-07	1976-77	2006-07	1976-77	2006-07
Paddy	638	874	1106	2131	2621	3868	Punjab	Punjab
Pigeon pea	933	691	600	650	1340	989	U.P.	Bihar
Groundnut	627	949	737	866	1106	1981	Haryana	Tamilnadu
Sorghum	682	1033	650	844	906	1033	Tamilnadu	Madhya Pradesh
Maize	1101	962	1033	1912	2617	3838	Karnataka	Tamilnadu
Pearl millet	641	1375	534	886	934	1649	Punjab	Haryana
Wheat	766	1916	1304	2708	3432	4232	Punjab	Haryana
Chickpea	530	988	683	845	301	1085	Punjab	Andhra Pradesh
Soybean	-	1063	-	1073	-	1515	-	Andhra Pradesh

2.5. PRESENT CROPPING PATTERN INAPPROPRIATE

The cropping pattern in the rainfed areas was generally determined by

a. Moisture availability period during which a particular crop or crop-sequence can be sustained,

b. Requirement of food depending on food habits and other domestic needs of the local population,

c. Requirement of raw material for local industry if existing in the locality.

Until recently 80 percent of the gross cropped area was under food crops, of which nearly 73 percent was under cereals and 27 percent under oilseeds, pulses and fiber crops. Present cropping pattern was determined largely by the average total annual rainfall in a particular locality, and under many conditions and at many places it was not appropriate, because average total rainfall of the locality alone did not determine the moisture availability period. This period was determined by rainfall distribution, slope of land and depth and texture of the soil.

A large proportion of wheat cultivation in the states was accounted for the farmers' thinking that growing of *rabi* crops might prove more rewarding than that of *kharif* crops. Consequently, comparatively less area was cropped in *kharif* and *kharif* - fallow - fallow - *rabi* or fallow - *rabi* rotations were followed over vast areas in the states. Since there were varying agro-climatic conditions within the same crop zone, paddy in *kharif* and wheat in *rabi* both suffered from moisture stress in certain areas. The recent developments in dryland technology and evolution of short duration high yielding and improved varieties of *kharif* crops by plant breeders had proved otherwise and as a result there is a desirable change in cropping pattern of the states (Verma, 1985).

2.6. LOW YIELDS OF WHEAT AND PADDY

The average yields of both rice and wheat in Madhya Pradesh and Chhattisgarh states were much below the national average. These are 638 and 766 kg ha^{-1} compared to the national average of 1108 and 1394 kg ha^{-1}, respectively. These low yields were mainly because the major portion of the areas under these crops was rainfed. Only 16 percent and 20 percent of the cropped areas under paddy and wheat, respectively, were irrigated in the states. Also, the irrigation partly depended on the intensity and the amount rainfall in the catchment area of various reservoirs. The yield of rainfed paddy was dependent upon the amount and distribution of rainfall during crop growth period, and the soil type on which it was grown. As it is much susceptible to drought than any other crop, no cultivator will take the risk of applying fertilizers and investing money on other improved practices recommended for this crop, unless assured water supply to the crop when required was available. If the rain did not occur in time, in spite of using the whole package of improved practices including plant protection measures, the farmer might get a poor yield or even no yield due to complete crop failure as a result of severe drought. This does not happen in case of upland crops like sorghum particularly when short duration and high yielding varieties are grown.

2.7. CHRONICALLY DROUGHT AFFECTED AREAS

According to Samra (2006) about 28 percent of geographical area of India was vulnerable to drought. The probability of occurrence of drought in Madhya Pradesh and Bihar states was once in 5 years, in eastern Uttar Pradesh, Chhattisgarh and Jharkhand once in 4 years, and in western Utter Pradesh once in 3 years.

A survey of the chronically drought affected areas disclosed that inappropriate cropping pattern in these areas was largely responsible for it. Is it not surprising that even Chhattisgarh state had some chronically drought affected areas although it received more than 100cm of average annual rainfall? The reason was rice being planted on all sorts of soils including shallow and light soils in which moisture availability period was low and even small dry spells between the rains resulted in either drying of the crop or in considerably reduced yields, depending upon the duration of dry spell.

Small pockets in districts of Jhabua and Dhar were frequently affected by drought although the average annual rainfall (less than 80 cm) was not so low. Steep land slope, undulating topography and light soils having low water holding capacity and predominance of maize crop on such soils, because maize was the common food stuff of the tribals of the area, were the main constraints responsible for drought in some pockets. Growing of maize crop further aggravated the bad situation as this crop was more susceptible to drought than sorghum. In Tikamgarh, Chhatarpur, Satna, Rewa and Damoh districts also drought occurred frequently due to light soils, growing long duration varieties of sorghum in *kharif* and rainfed wheat grown in *rabi* on profile stored moisture which was rather limited due to light texture of the soil.

In Mandla district, drought affected rice crop as the soils of this area were mostly skeletal on steep slopes. The crop was affected due to drought as soon as there was wide gap in rains. In Jabalpur, Shahdol, Sidhi, Bilaspur, Raipur and Durg districts again it was rice crop that was adversely affected by chronic drought in patches with light soils and undulated fields. As distribution of rainfall was mostly erratic and wide gaps in rainfall were frequent, the drought was imminent as no effort was made to store the runoff water and recycle it for at least the life saving irrigation to the rice crop.

2.8. STRATEGY OF RAINFED FARMING DEVELOPMENT

Before deciding the strategy of rainfed farming development, an examination of the constraints faced was necessary. The major problems of rainfed farming have been discussed already but various constraints in detail are presented as follows (Verma, 1980a):

2.8.1. Constraints

To summarize, the productivity of Rainfed Farming suffers from following constraints:

a. Climatic constraints

b. Land/soil constraints

c. Cropping system constraints

d. Farming system constraints

2.8.1.1. Climatic constraints : Rainfed farming was practiced under various climatic zones *viz.* arid, semi-arid, sub-humid and humid zones. There is large variation in rainfall, onset and withdrawal of monsoon within these zones and from zone to zone. Various climatic constraints are:

i. **Late onset of monsoon:** Late onset of monsoon influenced the sowing time and consequently the productivity of crops.

ii. **Early withdrawal of monsoon:** Early withdrawal of monsoon resulted in reducing the moisture availability period and consequent mild to severe drought. The *kharif* crops were, accordingly subjected to moisture stress resulting in poor yields. *Rabi* crops grown on soil-profile stored moisture may not be sown at all or fail to mature due to inadequate soil moisture, particularly during flowering and maturity stages. Thus, early withdrawal of monsoon had equally or more adverse effect on crop-yields than late onset of monsoon.

iii. **Variation in quantity of rainfall:** The quantity of rainfall received varies from year to year affecting rainfed crops by drought or waterlogging.

iv. **Variation in distribution of rainfall:** Distribution of rainfall over time (*i.e.* number of rainy days) was more important than variation in total annual rainfall. Less than normal rainfall but well distribution over the season was better than greater amount of rainfall through fewer intense showers, which caused soil erosion, waterlogging and temporary flooding on one hand intermittent dry spells on the other if there was a big gap between two intense showers.

v. **Untimely rains:** Untimely rains may adversely affect the crops while cloudy weather could invite pests and diseases to harm the *rabi* crops.

vi. **Unsuitable temperature regime during crop-growth:** Low temperature leads to frost attack while high temperature or hot wind at dough stage hinders proper seed formation and development and produces thin and small size seeds.

2.8.1.2. Land/Soil constraints

2.8.1.2.1. Topography of land: Moderate slope of land (one to two percent) was desirable for raising crops on various types of soil. Steep slope of land (slope >10 percent) was a big constraint because

1. The land was very susceptible to water erosion. The run off gained high velocity soon after it started flowing on a slopy land and consequently the run-off became highly erosive. Steep slope was not conducive to undertake ploughing, seeding and intercultural operations against the slope to conserve soil and water. According to Universal Soil Loss Equation (USLE), soil losses of 500 tonnes ha^{-1} or more were estimated for slopy land. The severity of such erosion can be judged from the fact that a soil loss of 100 tonnes ha^{-1} will remove the whole top soil in 20 years (Ayres 1936).
2. The upper part of slopy land does not absorb much rain water and dries fast after rains. Thus moisture availability on slopy land was less, and vegetation cover was sparse or nil. Hillocks might be sliding down at top while lower part was stable due to vegetative cover. Soil slips, slides, glides and mudflows together categorized as 'mass movement'

are common in areas of high intense rainfall, particularly where permeable soil rested on steep slopes (Karale *et al.,* 1991).

3. The soil of slopy land suffered from excessive leaching resulting in impoverishment of the surface soil which could be deprived of finer particles also.
4. Steep land usually has a soil shallow in depth and coarse in texture and consequently less suitable for raising crops particularly under rainfed conditions.
5. Steep slope and arid climate was the worst combination for land, leading to a very fragile system.
6. Adverse effects of drought were aggravated on steep land.

2.8.1.2.2. Flat topography: Flat topography of land particularly in case of Vertisols was also a constraint because it resulted in waterlogging/water stagnation which led to failure of most *kharif* crops. Saucer shaped topography of land was responsible for the lack of external drainage of land and resulted in problems of soil salinity and sodicity unfavourable for crop growth.

2.8.1.2.3. Soil Depth: Soil depth was a big constraint for arable farming, particularly under rainfed conditions. Shallow soil was invariably found on steep hill slopes and was deprived of fine particles and nutrients resulting in low productivity. In Bihar, the upland soils-red and laterites, are shallow to medium deep, coarse textured with low water holding capacity. Consequently the soils have severe limitation for rainfed farming. In Jharkhand, hilly tracts of Chhotanagpur Plateau also have very shallow soil.

2.8.1.2.4. Physical properties

i. **Texture:** Sandy and Sandy loam soils have high infiltration rate and are too permeable to retain much rain water in soil profiles. Consequently their productivity under rainfed conditions was low.
ii. **Crusting:** Some soils contain critical ratio of sand and silt leading to common problem of soil-crusting, particularly in low rainfall areas, lack of organic matter and high percentage of fine silt are the factors contributing to crusting of soil *i.e.* sealing of surface soil by fine particles. Soil crusting results in poor germination of seeds, if there is a rain immediately after sowing.

A simple index for crusting can be calculated as follows (Karale et al 1991):

$$\text{Index of crusting} = \frac{Zf + Zc}{C}$$

where Zf = percent fine silt

Zc = percent coarse silt

C = percent clay

The index is 1.5 for crusting soil and 2.5 for soil with intense crusting. The index can be calculated based on organic matter content also as below:

$$\text{Index of Crusting} = \frac{1.5\,\text{Zf} + 0.75\,\text{Zc}}{\text{C} + 10\,\text{O.M.}}$$

where O.M. = per cent organic matter

The value of this index varies from 0.2 for non-crusting soil to 2.0 for soil with very intense crusting.

iii. **High bulk density**: Some soils develop a high bulk density layer below surface soil due to illuviation or as plough sole. This soil layer has a high mechanical impedance which restricts root growth and reduces the crop yields in such soils. In such cases breaking of the hard layer by sub-soiling/chiseling increased crop yields. Many red soils containing 20 to 40 percent clay and silt, *i.e.* sandy loam texture get hard on drying, have low moisture storage capacity and dry up quickly. The dry solid mass has a high mechanical impedance due to high bulk density and root growth of most *kharif* crops suffers with consequent low yields.

iv. **Low infiltration and poor internal drainage:** Vertisols have a high clay content and dominance of montmorillonitic type of clay minerals. Consequently these exhibit enormous swelling on wetting and shrinking on drying. When saturated with water, these become almost impervious and result in low terminal infiltration rates and poor internal drainage. Also vertical heterogeneity of hydraulic conductivity of such soils indicated occurrence of less permeable layers underlain by more permeable ones in the profile (Kauraw and Gupta 1985) and this situation provided an ideal condition for the development of pseudo/perched water-table after the rains. Due to this waterlogging/water stagnation often takes place during rainy season particularly when the topography is flat. Poor internal drainage and waterlogging/water stagnation are the main constraints to successful crop production on Vertisols in rainy season.

v. **Narrow range of moisture content for workable condition:** A big constraint of Vertisols is that these can be operated upon during a narrow range of moisture content. When wet, these are sticky and plastic and therefore not workable. Also, on drying these become too hard to operate upon to prepare a seed-bed. After wetting of surface soil by rains a few dry days are required to reduce the soil moisture content to a level at which the soil is workable. Sometimes it becomes difficult or impossible to sow a *kharif* crop due to lack of dry days and area under *kharif* crops was reduced. Further, if the dry period during the rains gets prolonged the soil becomes too hard to work on it and again sowing many not be possible. This means that only a few days are available for sowing *kharif* crops on Vertisol, once the monsoon starts. Similarly, after the harvest of *kharif* crop, sowing of *rabi* crop should be done as soon as possible, otherwise soil moisture of surface soil may recede to a level at which soil is not workable (Sharma et al 1983).

vi. **Greater moisture loss through shrinkage cracks:** When post-rainy season crops growing on Vertisol deplete the soil moisture from surface soil through evapo-transpiration, shrinkage cracks are formed and consequently much greater soil surface (3.0 to 4.6 times) was exposed to evaporation. The additional water loss through evaporation from cracks is found to be as high as 28 percent of that evaporated at the surface, in a study conducted at Indore (Madhya Pradesh) by Sharma et al (1982). This loss of moisture reduces yield of crops, because the profile stored moisture was limited.

2.8.1.2.5. Chemical properties

i. **Soil salinity:** Saline soils under rainfed conditions are formed on a piece of land with restricted external drainage. The run-off from uplands around accumulates on the land and gets evaporated after monsoon rains cease. This goes on for years and the saline soil is formed as a result of continuous concentration of salts brought by run-off. This soil usually contains chlorides and sulphates of sodium and calcium which are water soluble.

ii. **Soil Sodicity:** Salt affected black soils formed on patches of varying dimensions are invariably saline-sodic. A saline sodic soil has much more unfavorable environment for plant roots as compared to saline soil. The already low permeable black soil is rendered almost impermeable at sodicity levels corresponding to exchangeable sodium percentage (ESP) greater than 22 (Gupta and Verma 1983). Water retention by the soil increases and water release decreases due to the loss of macro-pores. Crops grown on such a soil either fail or yield very low due to: (a) poor soil tilth, aeration and root development, (b) poor availability of nutrients and (c) nutritional disorders caused by excess of Na^+ (high pH).

iii. **Soil acidity:** Heavily weathered lateritie and lateritic soils have low base saturation and are consequently acidic (pH below 6.5). Acidic soils are usually low in calcium and phosphorus, get rich in Fe and Al. Red acidic soils of Netarhat, Bihar, contained more exchangeable Al^{+++}. Availability of Mo and B decreases with increase in acidity. The adverse effect of acidity on legumes was due to Mo deficiency. The acid soils invariably contain toxic amounts of Al^{+++}, Fe^{+++} and Mn^{++}. But sedentary acid soils of Bihar and Jharkhand which are sandy and highly leached suffer from Zn deficiency.

2.8.1.2.6. Biological properties: Soils poor in organic matter are also poor in micro flora and fauna - bacteria, actinomycetes, algae, fungi and protozoa. These micro organisms play a very important role in soil productivity through various processes such as nitrogen fixation, mineralization of organic nitrogen, solubilisation of soil – phosphorus, decomposition of plant and animal residues, humification, etc (Kauraw and Verma 1982).

2.8.1.3. Appropriate cropping system: Ignorance about the appropriate cropping system for a particular farmer's holding was a big constraint in the development of rain-fed farming. Leaving the land fallow during rainy season and raising a crop on profile stored moisture was an undesirable practice as profile stored moisture was limited and response to other inputs like good seed, fertilizer etc. was also limited due to inadequate soil moisture. Amongst *kharif* crops also, more remunerative crops befitting his land and soil i.e. moisture availability period, are not being adopted.

2.8.1.4. Farming System: Not only a cropping system for every agricultural situation, but also an appropriate farming system *viz.* crop-dairy, crop-fisheries, crop-poultry, etc. must be adopted. In humid regions rice-fish system was very successful. For slopy land with shallow soil, alternate land-use rather than raising a crop, was more profitable. That is, on such soil horti-pastoral or silvi-pastoral system will not only be more remunerative but the soil will also be conserved against water-erosion.

2.8.2. Strategy

One of the most important reasons for low crop-yields was lack of availability of adequate moisture throughout the growth period of crops as irrigation facility was not available and proper management of rain-water was not adopted. Apparently the strategy of rainfed farming development should be (Verma, 1988):

i. **Proper management of rain-water on watershed basis**: In semi-arid and sub-humid areas, lack of adequate moisture throughout the crop-growth period was often felt. There were problems of water-erosion on slopy land and of waterlogging/water stagnation on flat land, particularly in black soil regions. Proper management of water will lead to maximum conservation of water in situ, safe disposal of excessive run-off water, storage of run-off in appropriate manner at suitable places and recycle it for life saving irrigation during dry spell or for a come-up irrigation for double cropping.

ii. **Appropriate cropping system:** An appropriate cropping system including more remunerative crops tailored to moisture availability period and suiting the land and soil resources, must be adopted, so that rainfed farming was not only more paying but was also sustainable. On slopy land with shallow soil, alternate land use rather than cropping must be adopted. Further, improved package of practices including fertilizer-use, good quality seed of appropriate variety, timely harvest at physiological maturity stage must be adopted. Also, improved implements must be used to increase the efficiency of farm labour.

iii. **Appropriate farming system**: Not only appropriate cropping system but an appropriate farming system must be adopted by all farmers, depending upon size of holding, type of land and soil, etc. Mixed farming (crop-dairy) was traditionally practiced by most farmers. While there was lot of scope to improve and make the crop-dairy system more profitable it was necessary to adopt other appropriate farming systems as well, so as to have a second string to farm-income. Other farming systems are agri-horti system, agro-forestry, goatary, piggery, rice-fish farming, bee-keeping, lac cultivation, etc.

■■■■

III

LAND AND WATER RESOURCES

Central India is characterized by endowment of rich and diverse natural resources. Land, water, climate and bio-inhabitants (human beings, livestock and plants) are the basic resources of a Country/State for its agricultural development. Of these land and water are the most important resources in reference to optimum crop-production which essentially needs appropriate quality and adequate quantity of these resources. Land and water resources are valuable gifts of the nature to a nation and the prosperity of the nation depends on how well these resources are utilized and managed. The importance of land and water resources to living beings cannot be over emphasized. Our ancenstors worshiped *jal devta* (water god) and *bhumi mata* (mother soil) as is evident from a quotation from Rigveda, which means as follows:

Soil is my mother and I am son of mother earth, water is my father, let both of them lead me to perfection.

In natural eco-system, these resources are in equilibrium but due to unprecedented increase in population of human beings and livestock and consequent high demand for food and fodder, the resources have been exploited indiscriminately leading to significant reduction in the productivity of land and water resources. A poor land is the habitat of poor people. History is full of examples of ruin of civilizations in Indus valley and even in Europe due to desertification and environmental degradation resulting from mismanagement of land and water resources. A quotation from USDA year book 1957 entitled "SOIL" also emphasized: "Stable civilizations are associated with long familiarity with a soil, they grow out of it and are rooted in it". An inventory of both these natural resources *viz.* land and water was therefore, necessary for efficient planning and management of agricultural production.

3.1. LAND RESOURCE

Land resource was the most important non-renewable natural resource of a country and was basic to agriculture. India is very rich in land (soil) resource. Out of its total geographical area of 329 million (M) ha, 142 M ha (43.5 percent) is cultivated, as against only 11.3 percent in the world. The Central India consisting of 5 states occupied geographical area of 85.82 M ha, which is 26.1 percent of the total geographic area of the country. Undivided Madhya Pradesh had the largest geographical area of all the states in the country that is 44.28 million ha. After formation of two separate states, the geographical area of Madhya Pradesh is 30.76 million ha (69.63 percent) and that of Chhattisgarh state 13.52 million ha (30.37 percent). Similarly, the geographical area of undivided Bihar was 17.3877 million ha while present Bihar state has a geographical area of 9.4163 million ha (54.15 percent) and Jharkhand state 7.9714 million ha (45.85 percent). Uttar Pradesh state has a geographical area of 24.928 million ha. Thus, all the five states are blessed with vast land surfaces enjoying varied agro-climatic conditions. Besides vast area of land resource, most of alluvial plains in Bihar and Uttar Pradesh, plains of central Chhattisgarh, Narmada Valley, Malwa Plateau, Kymore Plateau and Nimar plains have very good land suitable for intensive cultivation.

3.1.1. Physiography

Central India displays a diversity of land-forms consisting of hilly regions, low hills, plateaus, plains, river-valleys and shallow to very deep ravines including world infamous Chambal ravines. Physiographically the Central India could be divided into many regions the most important five of which are:

i. Indogangetic plains (River valleys including low lying plains)
ii. Central and south-eastern plateaus (Malwa, Kymore, Bastar and Chhotanagpur Plateaus)
iii. Vindhyan, Satpura and Maikal stretch,
iv. Narmada Valley, and
v. Chhattisgarh plains

These regions covered diverse areas ranging in elevation from less than 50 metres to more than 1200 metres above mean sea level. The vast stretch of Indo-Gangetic plain covers most of Uttar Pradesh and Bihar states. Bihar and Eastern Uttar Pradesh have large areas under *Tal, Diara* and *Chaurs* lands, inhabiting a sizeable population fully dependent on agriculture (Singh and Diwakar, 1988, Chandra, 2006). The *Tal* land in Bihar extends from Lakhisarai to Fatuah over a length of about 105 km and covers an area of about 1.04 lakh ha. *Tal* land is low lying flat area having no external drainage. Consequently during monsoon season, rain-water gets accumulated up to a depth of about five metres, and the whole area is inundated for a period of more than four months in a year, i.e. during whole of the *kharif* season. The *Diara* land in Bihar covers an area of about 11.6 lakh ha. The land has an undulating topography with varying patterns of low land and upland. It also occurs in flood plains of eastern Uttar Pradesh in continuation to that in Bihar. The *Chaurs* is a large tract of land in North Bihar having a fragile and instable eco-system. It remains waterlogged for a variable time and about four lakh ha is inundated during different periods in a year.

The Western part of Indo-gangetic plain, drained by Yamuna, Ganges (Ganga) and Gomati rivers, comprised of *Terai, Bangar* (upland formed of old alluvium) and *Khadar* (lowland covered by recent alluvium). *Terai* was sub-

Himalayan land and *Khadar* was riverine and flooded by rivers almost every year. The Vindhyan range runs through middle of Madhya Pradesh and its northern slope drains into Yamuna through Chambal and its tributaries and its other tributaries *viz. Sindh, Betwa* and *Ken.* The southern slope of Vindhyan range drains into *Narmada.* A small part of south western Madhya Pradesh was drained into *Mahi* river. North-western part of Madhya Pradesh had very fragile land carved into deep to very deep ravines by *Chambal* river. *Betwa* river also formed shallow to deep ravines.

3.1.2. Land use pattern

In undivided Madhya Pradesh more than 43 percent (192.06 lakh ha) of the total area was cultivated, and less than 7 percent (29.42 lakh ha) was under permanent pastures, grazing land, orchards and tree crops. While about 8 percent (34.44 lakh ha) of the total land was fallow and culturable waste, approximately 32 percent (140.5 lakh ha) was under forests. Thus, only about 10 percent of the total land (45.68 lakh ha) was not available for agricultural and forestry use, and the remaining 90 percent of the land (about 395 lakh ha) was available for agricultural and allied activities. Land use pattern of Madhya Pradesh and Chhattisgarh states (2006-07) is presented in table 4 (Department of Agriculture 2009, DFW&AD, 2009).

Table 4. Land use pattern of Madhya Pradesh and Chhattisgarh states (lakh ha) (2006-07).

S.No.	Land use	Madhya Pradesh	Chhattisgarh
1	Forest	85.95 (27.9)*	63.03 (46.6)
2	Land not available for cultivation	33.96 (11.0)	10.13 (7.4)
2.1	Land put to non-agricultural use	19.91 (6.45)	
2.2	Barren and uncultivable waste	14.05 (4.55)	
3	Other uncultivated land excluding fallow	13.67 (4.4)	8.53 (6.2)
3.1	Permanent pastures, grazing land	13.48 (4.34)	8.53 (6.2)
3.2	Land tree crops, groves	0.19 (0.66)	Nil
4	Cultivable waste land	11.77 (3.8)	3.36 (2.4)
4.1	Land that can be cultivated immediately	5.36 (1.73)	
4.2	Land cultivable after improvement	3.59 (1.16)	
4.3	Uneconomical patches of land	2.82 (0.91)	
5	Fallow land	13.81 (4.5)	5.19 (3.8)
5.1	Current	7.69 (2.51)	2.80 (2.0)
5.2	Old	6.12 (1.99)	2.39 (1.8)
6	Net area sown	148.4 (48.2)	47.6 (34.5)
7	Area sown more than once	53.8	5.64
8	Gross cropped area	202.2	53.27
9	Cropping Intensity	136 percent	134 percent

*Figures in parenthesis show percentage to total area

Of about 30.8 million ha land in Madhya Pradesh, 14.8 million ha (48.2 percent) is cultivated and only 8.6 million ha (27.9 percent) is under forests, while in Chhattisgarh, 4.8 million ha (34.5 percent) out of total land of 13.5 million ha is cultivated and as much as 6.3 million ha (46.6 percent) is under forest. But in Madhya Pradesh land not available for cultivation or afforestation is 3.4 million ha (11 percent) which is much more than that in Chhattisgarh - a little more than 1 million ha (7.4 percent). Also cultivable waste land is much higher (11.8 lakh ha *i.e.* 3.8 percent of total land) in Madhya Pradesh than in Chhattisgarh (3.4 lakh ha *i.e.* 2.4 percent of total land). Fortunately a major part of it can be either cultivated immediately (5.36 lakh ha) or after some improvement (3.59 lakh ha). Only the remaining 2.82 lakh ha land is difficult to reclaim but it may not be uneconomical to reclaim in view of the ever escalating land prices and alternate land-use technology now available. It is interesting to note that the magnitude of fallow land, particularly current fallow, is higher in Madhya Pradesh as compared to Chhattisgarh. Thus, there is greater potential to increase cropped area and consequent higher crop-production in Madhya Pradesh.

3.1.3. Forest cover

Tropical and sub-tropical forests in Central India are very rich in different kind of plant species including multipurpose trees, fruit plants, shrubs, bamboo, canes and beautiful flowering creepers and medicinal plants. The forest products are a source of livelihood for the people, particularly the tribals. As shown in table 5, total forest area in Central India is about 19.14 million ha which is 22.3 percent of the total geographical area of the region, and is slightly higher than national average (20.64 percent). The area also constitutes about 28 percent of the total forest area (67.8 million ha) of the nation. Of all the states of central India, Chhattisgarh has the largest percentage (46.6 percent) of its geographical area under forest, followed by Jharkhand and Madhya Pradesh. Both Uttar Pradesh and Bihar states have only about 6 percent of their geographical area under forest and thus the lowest forest cover in the region. Madhya Pradesh contributes the highest (44.9 percent) towards total forest area of the region, followed by Chhattisgarh (32.9 percent) and Jharkhand (11.9 percent). The vegetative cover in Uttar Pradesh and Bihar has been over-exploited and concerted efforts are necessary to restore forest cover to match the national level at least. This target can be easily met by suitable plantation on bare hill slopes, barren lands, etc.

Table 5. Area under forest in Central India (lakh ha)

States	Geographical area	Forest area	Forest area as % of geographical area
Bihar	94.163	5.558	5.9
Jharkhand	79.714	22.716	28.5
Madhya Pradesh	307.610	85.950	27.9
Chhatisgarh	135.190	63.030	46.6
Uttar Pradesh	240.928	14.118	5.9
Total	857.605	191.372	22.3

3.1.4. Land degradation

The land in Central India is under immense pressure for intensive agriculture particularly in Bihar and Uttar Pradesh as these are most densely populated states in the Country. As per census 2001, Bihar is the most populous state with a population density of 881 persons per km^2 against the national average density of 324 persons per km^2. The next most populous state is Uttar Pradesh with 690 persons per km^2. Besides intensive agriculture, industrialization, urbanization and development of roads and other infrastructure take a heavy toll on good agricultural land. Further, unsystematic exploitation of other natural resources through excessive mining, reckless and rapid deforestation and extensive use of class vi and vii land and even hill slopes (through terracing) for raising upland crops have caused widespread land-degradation through serious soil erosion not only by formation of rills but also through gullies and deep ravines (up to 26 metres deep). High rainfall and farmer's practice of leaving the land fallow in rainy season aggravate the soil loss. According to Maji (2007) of 120.72 million ha of degraded land in India, water erosion alone accounts for 82.57 M ha (68.4 percent) followed by chemical degradation (formation of saline, alkali and saline-alkali soils) 24.68 M ha (20.4 percent) and wind erosion 12.4 M ha (10.3 percent). As regards the erosion rate of different land resource regions, it is the highest on alluvial soils in Central India followed by gullied black soil. These erosion rates are much higher than those in *Shiwalik* region (80 t ha^{-1}), north eastern region with shifting cultivation (27-40 t ha^{-1}) and northern Himalayan forest region (2.1 t ha^{-1}) – the least in the Country.

As regards Central India states, the land surface is mostly undulating particularly in Jharkhand, Uttar Pradesh, northern Chhattisgarh and Madhya Pradesh except in Gangetic plains and Narmada Valley, and is subject to moderate to severe water erosion. The *Chambal* valley in north western Madhya Pradesh has severe land degradation by formation of deep to very deep ravines on alluvial soils. Western part of Jharkhand, north eastern and southern part of Chhattisgarh, Bundelkhand region of Uttar Pradesh and Madhya Pradesh and Jhabua region of Madhya Pradesh have moderate to severe limitations of land-type for raising good crops.

3.1.5. Extent of degraded land

Many workers (Sehgal and Abrol 1994, Bali 1974) have attempted to determine the extent of degraded land in Central India. While all degraded lands are waste-land, all waste lands are not necessarily degraded. Wasteland has been defined as uncultivated non-forest land. Mathur (2009) reviewed the work on assessment of waste land/degraded lands in Central India using remote sensing and GIS techniques and concluded that Uttar Pradesh and Madhya Pradesh had 14.58 and 14.0 million ha, respectively, of waste land/degraded land while Bihar had the least (1.46 M ha). As already stated, due to severe land degradation and diversion of good agricultural land to industrialization, urbanization, and build up of infrastructure (roads, railway lines, office buildings, etc.) on one hand and incessantly increasing population on the other, the per capita available agricultural land has been reduced to 0.44 ha (0.31 ha crop land and 0.13 pasture land) in 2005-06 (Sharda, 2009) and the same is projected to be 0.121 ha in 2025 and 0.087 by 2050 when the population stabilizes. Thus, the need to halt land degradation is of top-priority for the survival of all living beings.

3.1.6. Soil Types

Central India has a variety of soils originating from different types of rocks transformed by different paedogenic processes, ranging from rich clayey to gravelly in nature and shallow to very deep in depth. Majority of the area is covered by four broad soil types, *viz.* Inceptisols (49 percent), Entisols (21 percent), Alfisols (18 percent) and Vertisols (11 percent). In Uttar Pradesh, nearly 70 percent area is under Inceptisols, 19 percent under Entisols, 5 percent under Alfisols, 2 percent under Vertisols and small area under Mollisols. In Bihar, Inceptisols, Alfisols, Entisols and Vertisols cover about 34.9, 34.5, 27.3 and 0.6 percent of geographical area (Sharda 2009). The major soil groups in Madhya Pradesh are alluvial, black (deep, medium and shallow), and mixed red and black. Chhattisgarh, Bihar and Uttar Pradesh states are comparatively more uniform and have only two main soil types, *viz.* red and yellow, and red and lateritic soils. The skeletal soils earlier considered as one of the soil types, have been grouped under mixed red and black soil (Tomar et al 1995, Velayutham and Bhattacharya 2000). Various soil types in Central India are discussed below, in brief.

3.1.6.1. Alluvial soil : The colour of these soils ranged from pale yellow to grayish brown. The texture of surface soil varied from sandy loam to clay loam. The middle horizon was richer in clay content due to illuviation of finer particles. The soils were neutral to alkaline in reaction and the calcium carbonate content increased with depth. In Madhya Pradesh besides northern plains, patches of alluvial soils are found in the beds of some rivers in Satna district, and on either side of *Wainganga, Bag* and *Dev* rivers in Balaghat district. Vast stretches of alluvial soil exist in Uttar Pradesh, Bihar and Central Chhattisgarh.

3.1.6.2. Black Soils: These soils were derived from trap and varied in texture from loam to clay. Montmorillonite type of clay mineral dominated in such soils and their characteristic features were 'gilgai' formation, closely intersecting slicken sides and development of shrinkage cracks. Depending upon their depth, these were again divided into three sub-groups:

1. **Deep black soil**: These were more than 90 cm in depth and dark grayish brown in colour. Their clay content ranged from 40 to 60 percent. According to taxonomic classification the soil is called Vertisol.
2. **Medium black soils:** These soils varied in depth from 30 to 90 cm and were usually dark brown in colour with some light coloured or reddish coloured patches. Clay content varied from 30 to 40 percent. According to taxonomic classification, these are classed as Vertic ustocrepts.

3. **Shallow black soils:** These were less than 30 cm deep and their texture was clay loam with 20 to 30 percent clay. These soils do not show the characteristic features like slickensides and gilgai micro-relief, and are accordingly classed as Vertic ustorthents.

3.1.6.3. Mixed red and black soils : Red and black soils occurred side by side, but in a topo-sequence *i.e.* on slopy land, reddish yellow soil found on top or ridge, red soil on mid-slope and finally black soil on the lowest part of the slope. These soils in topo-sequence are locally called *Ranker* (light and gravelly soil), *Padua* (medium soil) and *Kabar* or *Mar* (heavy black soil). Red soils were coarse textured and devoid of lime concretions and free carbonates. Kaolinite was the predominant clay mineral in these. Black soils were comparatively less dark in colour and contained intergrades of montmorillonite, illite and kaolinite type of clay minerals. In taxonomic classification, the soils are named as ustorthents, ustifluvents and ustocrepts depending upon differentiation of horizons.

3.1.6.4. Red and yellow soils : In this group, red and yellow soils occurred together on undulating plains. Yellow colour was due to higher degree of hydration of ferric oxide than in red soils. Texture varied from silty loam to silty clay loam. The soils were poor in calcium but iron concretions were usually present in lower horizons. These soils have been formed from a variety of rocks as micaceous quartzite, schists, hornblende and gneiss. In this group, four main types were distinguished locally, based on colour and texture: (1) *Kanhar* (dark clay soil with 40 - 60 per cent clay), (2) *Dorsa* (yellow loam with 20-30 per cent clay), (3) *Matasi* (yellowish brown sandy loam), and (4) *Bhata* (reddish yellow gravelly sandy soil). All these soil types usually occurred on a topo-sequence, with *Bhata* on top, followed by *Matasi* and *Dorsa* and finally *Kanhar* on the lowest part. Under taxonomic classification, these are Alfisol and Ultisol.

3.1.6.5. Red and lateritic soils : Laterites are highly ferruginous and apparently unstratified deposits, developed under high rainfall monsoon type of climate, *i.e.* tropical and sub tropical climate from basic parent rocks. The silicious matter of the rock is leached away almost completely during weathering and the sesquioxides (Fe_2O_3 + Al_2O_3) are left as a dominant content of the soil. The term laterite has been derived from latin word "later" which means brick. The laterites have reddish colour with maximum intensity in B-horizon which is rich in clay formed in situ and not migrated from A horizon. As laterites do not contain secondary clay minerals, these show the typical properties of clay (plasticity, cohesion, shrinkage, expansion, base exchange) to a small extent only. Red soils are derived from granites and gneisses and other rock formations such as sandstones, chernokites, diorites, hornblende, mica schists, quartzite and shales. On hill tops, these are very shallow and gravelly in nature and are called "*barra*" locally. In valleys, there are patches of good and deep red soils and even black soils. Taxonomic classification categorizes these into Oxisol, Ultisol and Alfisol classes.

3.1.7. Distribution of soils

Distribution of different soil types, their taxonomenclature and important properties were presented in table 6. In Bihar and Jharkhand (Chhotanagpur plateau) Vertisol (Swell-shrink soil) occurs mostly in alluvial tract (*Tal* land) and in piedmonts, toe slopes and valley flats in small patches (Singh and Diwakar 1988). Four soil series viz. Biru, Jarmundi, Kasiachak and Techatoli are identified under Vertisol (Karale *et al.,* 1988).

Table 6. Distribution, taxonomy and nature of various soil types

S.No.	Soil Type	Taxonomic class	Area (M ha)	Soil pH	Texture	$CaCO_3$ (%)	CEC Cmol p+ kg^{-1}
Madhya Pradesh							
1	Alluvial	Entisol, Inceptisol, Alfisol	3.35	Neutral to alkaline (7.0 to 8.5)	Sandy clay loam to Clay loam	1.0 to 4.2	17.7 to 28.3
2	Shallow & Medium black	Vertic ustorthents and Vertic ustocrepts	3.06	More or less neutral (7.0 to 7.5)	Clay loam to Clay	1.5 to 2.5	30.0 to 32.0
3	Medium & Deep black	Haplusterts	16.21	Neutral to alkaline (7.0 to 8.5)	Clay 40 to 60 percent	Calcareous	33 to 55
4	Mixed red & black	Ustorthents Ustifluvents Ustochrepts	8.11	7.1 to 8.9	Sandy loam to sandy clay	No lime concretions 0.5 to 0.8	12 to 34
5	Rock out- crops	–	0.09	–	–	–	–
Chhattisgarh							
1	Red and Yellow soils	Alfisol, Ultisol	9.70	6.8 to 7.6	Loamy sandy to clay	0.5 to 1.0	8 to 12
2	Red & lateritic soils	Oxisol Plenthaqults Plenthustalfs	3.91	5.6 to 6.2	Sandy loam to sandy clay	No $CaCO_3$	9.5 to 10.5
3	Rock out- crops	–	0.04	–	–	–	–
Bihar and Jharkhand							
1	Alluvial	Entisol, Inceptisol, Alfisol	16.79	7.0 to 8.5	Sandy loam to clay loam	1.0 to 4.0	16 to 30
2	Black soil	Vertisol, Vertic ustorthents	0.11	7.0 to 8.0	Clay loam to clay	1.5 to 4.5	30 to 32
3	Rock out- crops	–	0.02	–	–	–	–
Uttar Pradesh and Uttarakhand							
1	Alluvial	Inceptisol	20.9	6.5 to 8.5	Loam to clay loam	1.0 to 3.5	20 to 30
		Entisol	5.8				
		Alfisol	1.6				
		Mollisol	0.1				
2	Deep black soil	Vertisol	0.5	7.0 to 8.0	Clay loam to clay	1.5 to 4.0	30 to 35
3	Rock out-crops	–	0.41	–	–	–	–

3.2. WATER RESOURCES

India is quite rich in water resources. Varying amounts of annual rainfall (<100 mm to about 2500 mm) is received in different parts of the country. Only about 30 percent part of the country receives an annual rainfall less than 500 mm while 70 percent gets more than 500 mm. Around 394 million ha m of water are delivered by rains to Indian soil every year (Khanna 1991, Katyal 1992), as average annual rainfall of the country is 1198 mm and geographical area 329 million ha and about 6 M ha m is snow-melt. Of this 400 M ha-m, 70 M ha m evaporates to atmosphere, 115 M ha m appears as surface run-off (only 77.86 M ha m according to Venkateswarlu 1991) and 215 M ha m soaks into soil, of which again 45 M ha m regenerates as surface run-off into live *nalas*, seasonal rivers and rivulets, etc., 125 M ha m is retained by soil as soil-moisture and 45 M ha m (45.6 M ha m according to Padmanabhan 1988) charges the underground water. The total amount of surface flow (run-off) is thus as follows:

115 M ha m from rainfall in India

20 M ha m from out side India (Nepal, Tibet, etc.)

45 M ha m regenerated from soaked water

----

180 M ha m Total run- off (surface flow)

Thus, 180 M ha m surface inflow available to the country should be a boon, but unfortunately very little utilization, and further, it is proving to be a curse as the land-degradation through water-erosion is caused by it. Of 180 M ha m, only 32 M ha m is being utilized at present as follows:

15 M ha m stored in big reservoirs of major irrigation projects

15 M ha m directly lifted from rivers, rivulets and live *nalas*

2 M ha m stored in tanks under minor irrigation projects

As shown above, only 17.8 percent of surface inflow is being utilized presently. On the whole, a total of 465 million ha-m of water resource including 400 M ha m rainfall, 20 M ha m outside run-off and 45 M ha m underground water is available, which comes to approximately 140 cms per ha of geographical area and 327cms per ha of cultivated area. It is really an abundant amount of water resource. But variation of rainfall in time and space makes its availability rather uncertain when needed and where needed. Like our country, Central India is also blessed with adequate water resources such as high annual rainfall (756 to 1562 mm) many perennial rivers and lot of underground water (15.2 M ha m). It has four major river systems viz. Ganga, Godavari, Narmada and Mahanandi contributing a total annual river flow of 85 M ha m.

3.2.1. Rainfall

Based on 50 years' rainfall data (Sahasrabudhe 1968) recorded at various Research Stations of Madhya Pradesh, the average annual rainfall of Madhya Pradesh state varied from 756 mm (at Mandsaur) in west to 1562 mm (at

Waraseoni) in east and from 822 mm (at Gwalior) in north to 1063 mm (at Chhindwara) in south (Table 7). Usually the monsoon rains are more common and intense during the months of July and August (main part of monsoon season) and much less in June and September, considering the 50-year average monthly rainfall data recorded at 11 Research Stations. In west M.P., Mandsaur received the minimum average annual rainfall (756 mm) followed by Jaora (798 mm), while maximum annual rainfall (909 mm) occurred at Indore followed by Khargone (868 mm) and Khandwa (866 mm). West Madhya Pradesh received a seasonal rainfall of 905 mm (97 percent of total annual rainfall), of which 620 mm (68 percent) fell in July and Aug. Also, in east Madhya Pradesh the minimum average annual rainfall (1063 mm) was received at Chhindwara, while maximum (1562 mm) at Waraseoni. Also, of total average annual rainfall of 1268 mm (for whole east Madhya Pradesh), 1137 mm (90 percent) was received in monsoon season (June to Sept.) and of this seasonal rainfall again, 783 mm (64 percent) was received in July and August months.

Table 7. Seasonal and Total Rainfall (mm) in Central India

Region	Situations	Months				Season Total	Annual Total
		June	July	August	September		
MP west	Gwalior	77	245	249	166	768	822
	Indore	134	280	232	174	820	909
	Jaora	99	267	231	149	746	798
	Khandwa	109	297	188	172	766	865
	Khargone	146	282	181	180	789	886
	Mandsaur	95	260	240	110	705	756
Average	1916-1965	110	272	226	160	768	839
	1966-2005	108	305	315	177	905	927
MP East	Sehore	171	519	390	252	1332	1412
	Powerkheda	131	425	384	223	1163	1265
	Rewa	105	394	357	181	1037	1168
	Jabalpur	157	483	417	211	1268	1431
	Chhindwara	160	321	227	179	887	1063
	Waraseoni	221	519	440	214	1394	1562
Average	1916-1965	145	428	355	209	1137	1268
	1966-2005	144	371	382	201	1097	1190
Chhattisgarh	Raipur	219	371	389	213	1192	1359
	Bilaspur	191	381	387	194	1153	1328
Average	1916-2005	210	424	405	207	1246	1416
	1966-2005	189	394	391	232	1206	1394
U. P. West	1966-2005	69	268	286	150	773	842
U. P. East	1966-2005	105	309	301	198	914	981
Bihar	1966-2005	175	348	298	227	1048	1219
Jharkhand	1966-2005	193	338	329	244	1105	1352

As regards Chhattisgarh, average annual rainfall varied from 1328 mm at Bilaspur to 1359 mm at Raipur. Of the annual rainfall, 1153 mm (89 percent) and 1192 mm (89 percent) was received in monsoon season at Bilaspur and

Raipur, respectively. Again of the total seasonal rainfall, 768 mm (67 percent) and 760 mm (64 percent) came during the months of July and August, at Bilaspur and Raipur stations, respectively. For the whole state of Chhattisgarh, the average annual rainfall was 1416 mm, of which 1246 mm (88 percent) was monsoon seasonal rain. Again of the seasonal rainfall, 829 mm (67 percent) was received in July and August months.

In Uttar Pradesh, eastern part of the state receives higher rainfall than western part and maximum rainfall occurs in July and August months. Similarly in Bihar maximum rainfall comes in July and August months and its average rainfall (1219mm) is higher than that of east Uttar Pradesh (981 mm). Jharkhand receives higher rainfall (1352 mm) than Bihar (1219 mm). As regards Central India as a whole, maximum rainfall is received in July and August and rainfall increases from west (756 mm in Mandsaur district) to east (1562 mm in Waraseoni of Balaghat district, Madhya Pradesh).

Table 8. Seasonal distribution of rainfall

S. No.	Station	Pre-monsoon (mm)	Monsoon (mm)	Post monsoon (mm)	Winter (mm)	Total (mm)	No. of rainy days
1	Gwalior	18 (1)	737 (46)	33 (2)	34 (2)	822	57
2	Indore	19 (4)	820 (67)	54 (4)	16 (3)	909	78
3	Jaora	12 (2)	746 (46)	26 (1)	14 (2)	798	51
4	Khandwa	15 (2)	766 (48)	65 (3)	19 (2)	865	65
5	Khargone	19 (2)	789 (39)	67 (3)	11 (2)	886	46
6	Mandsaur	15 (2)	705 (51)	25 (2)	11 (2)	756	57
7	Dhar	26 (4)	745 (58)	45 (3)	28 (5)	844	70
8	Jhabua	20 (3)	700 (51)	21 (2)	25 (4)	766	60
West M.P.		20 (3)	752 (61)	47 (2)	20 (3)	839	59
1	Sehore	15 (2)	1332 (55)	45 (1)	19 (2)	1411	60
2	Powerkheda	21 (3)	1163 (60)	50 (3)	11 (2)	1265	68
3	Rewa	32 (3)	1037 (62)	44 (1)	55 (3)	1168	69
4	Jabalpur	40 (8)	1268 (64)	65 (3)	58 (3)	1431	78
5	Chhindwara	45 (4)	887 (61)	71 (3)	60 (6)	1063	74
6	Waraseoni	43 (3)	1394 (67)	74 (4)	51 (5)	1562	79
East M.P.		30 (2)	1138 (62)	55 (1)	41 (3)	1264	70
1	Raipur	55 (12)	1192 (73)	67 (3)	45 (6)	1359	94
2	Bilaspur	57 (7)	1153 (64)	65 (3)	53 (5)	1328	79
Chhattisgarh		52 (7)	1245 (68)	69 (4)	50 (5)	1416	84

Figures in parenthesis indicate number of rainy days

3.2.2. Seasonal distribution of rainfall

As already stated both in Madhya Pradesh and Chhattisgarh states, major part of the annual rainfall is confined to monsoon season (Table 8). However, a small fraction of the total rainfall is received as pre- and post-monsoon rain. In west Madhya Pradesh, pre-monsoon rainfall varies from 12 mm (less than 2 percent of total rainfall) at Jaora to 26 mm (about 3 percent of total rainfall) at Dhar. In east Madhya Pradesh, the same varies from 15 mm (about 1 percent of total) at Sehore to 45 mm (about 4 percent of total) at Chhindwara. For whole of west Madhya Pradesh, average pre-monsoon rainfall is 20 mm (2.4 percent of total) while for east Madhya Pradesh, it is 30 mm (2.4 percent of total).

As regards post monsoon rainfall, it varies annually from 21 mm (at Jhabua) to 67 mm (at Khargone) in west Madhya Pradesh, from 44 mm (at Rewa) to 74 mm (at Waraseoni) in east Madhya Pradesh and from 65 mm (at Bilaspur) to 67 mm (at Raipur) in Chhattisgarh state. The annual averages for west Madhya Pradesh, east Madhya Pradesh and Chhattisgarh state are 839, 1264 and 1416 mm, respectively.

3.2.3. Winter Rainfall

A small fraction of total annual rainfall is received as winter rainfall also. In west Madhya Pradesh, it varies from 11 mm (at Khargone and Mandsaur) to 34 mm (at Gwalior), in east M.P. it varies with an average value of 20 mm (at Powarkheda) to 60 mm (at Chhindwara) and has a mean value of 41 mm for whole east Madhya Pradesh. Similarly in Chhattisgarh state the range for winter-rainfall is from 45 mm (at Raipur) to 53 mm (at Bilaspur) with an average value 50 mm for the whole state.

3.2.4. Number of rainy days

Distribution of rainfall over time (*i.e.* number of days) is more important than the total amount of annual rainfall received at a place, from the view point of growth and development of rainfed crops. Greater the spread of rainfall, better will be the availability of soil moisture to crop plants. Again, the number of rainy days every year varies considerably (Table 8). It is 46 (at Khargone) to 78 (at Indore) days annually with an average 59 days for whole west Madhya Pradesh, 60 (at Sehore) to 78 days (at Jabalpur) a year with a mean of 70 days for east Madhya Pradesh and 79 to 94 days annually with an average of 84 days a year for Chhattisgarh state. Surprisingly, there is a greater variation in number of rainy days where annual rainfall is comparatively less, and this makes the rainfall more effective on the whole when it is comparatively less in quantity

The break-up of annual rainy days into pre-monsoon, post-monsoon and winter seasons (Table 8) shows that rainy days are largely concentrated in monsoon season. For example, of 59 rainy days of the year in west Madhya Pradesh, 51 were in monsoon season and only 3 in winter season and of 70 rainy days of East Madhya Pradesh, 62 were in monsoon season and 3 in winter season. Similarly, of 84 rainy days of Chhattisgarh state, 68 days were in monsoon season, 5 days in winter season, 7 days in pre-monsoon and 4 days in post–monsoon season. Thus, Chhattisgarh state appears to have better distribution of rainfall, particularly during pre-monsoon, post-monsoon and winter season.

3.2.5. Rainfall in the past *vs.* recent past

Fortunately long range rainfall data were available for about 90 years (1916-2005) for various research stations in Madhya Pradesh and Chhattisgarh states. These were processed for monsoonal (month-wise) and annual total for two periods *viz.* 50 years (1916-65) and 40 year (1966-2005) as shown in table 9. Interestingly, both monsoon and annual rainfall in West Madhya Pradesh has increased slightly in recent past (1966-05) as compared to past (1916-65). Also, the increase is mainly due to higher rainfall in July and August months. On the other hand in east Madhya Pradesh, there is a slight decrease in annual and monsoonal rainfall recently, and this decrease is accounted for by reduced rainfall in July month. In Chhattisgarh state there is no significant change in monsoonal and annual rainfall during the past and the recent past, *i.e.* there is no change in average rainfall (both annual and monsoonal) with time. However, it must be noted that this information is based on long term average and year to year variation in rainfall at different place may be much more than these differences in long term average.

Table 9. Monsoonal and Annual Rain (mm) During 1916-65 and 1966-2005

Region	Year	Month				Total	
		June	July	Aug.	September	Monsoonal	Annual
Western M.P.	1916-1965	110	270	225	159	752	839
	1966-2005	108	305	315	177	904	927
Eastern M.P.	1916-1965	145	428	355	209	1137	1268
	1966-2005	144	371	382	201	1097	1190
Chhattisgarh	1916-1965	210	424	405	207	1245	1416
	1966-2005	189	394	391	232	1206	1394

3.2.6. Characteristics of rainfall at Indore

Verma and Robertson (1973) studied the characteristics of rainfall at Indore by analyzing 47 years (1924-1970) rainfall data recorded at Indore. The annual total rainfall varied from 429 mm in 1925 to 1499 mm in 1961 with an average of 979 mm for the period 1924 to 1970. The rainfall varied considerably from month to month and year to year. Based on average figures for 47 years, February was the driest month followed by April, March, January, December, May, November and October. Most of the rainfall (about 90% of the total) is received during June to September (Monsoon season). July is the wettest month (291 mm average monthly rainfall) followed by August (252 mm), September (218 mm) and June (127 mm). Monsoon starts in June or early July. Except September which showed the maximum variation in monthly rainfall, months with high rainfall also had, in general, the high variation of monthly rainfall from year to year.

3.2.7. Probability of Rainfall

An analysis of long term rainfall data to determine initial and conditional probabilities of weakly rainfall of at least 5, 10 and 20 mm during pre-monsoon,

monsoon, post-monsoon and winter periods for 77 locations in India has been conducted by Virmani *et al* (1982). The relevant data from this study presented in table 10 show that the probability of onset of monsoon is high during 3rd week of June at Indore, Jabalpur and Raipur and during last week of June at Jhabua and Rewa. Monsoon lasts up to first week of September at Jhabua, upto 2nd week of September at Indore, Rewa and Jabalpur, and upto 3rd week of September at Raipur. Also, there is a high probability of post-monsoon rainfall ≥ 5 mm during the last week of September and first week of October at Raipur.

Table 10. Probability of Rainfall at Important Stations

Standard week	Indore		Jhabua		Jabalpur		Rewa		Raipur	
	≥ 5 mm	≥ 20 mm	≥ 5 mm	≥ 20 mm	≥ 5 mm	≥ 20 mm	≥ 5 mm	≥ 20 mm	≥ 5 mm	≥ 20 mm
21	0.17	0.00	0.07	0.01	0.13	0.06	0.13	0.00	0.25	0.06
22	0.27	0.17	0.18	0.10	0.25	0.14	0.21	0.00	0.31	0.12
23	0.43	0.13	0.31	0.26	0.43	0.16	0.38	0.21	0.44	0.29
24	0.77	0.47	0.53	0.37	0.74	0.43	0.46	0.29	0.84	0.54
25	0.70	0.53	0.62	0.46	0.78	0.57	0.67	0.33	0.94	0.75
26	0.83	0.60	0.79	0.62	0.90	0.78	0.88	0.58	0.96	0.84
27	0.97	0.83	0.87	0.66	0.99	0.84	0.96	0.83	1.00	0.88
28	0.90	0.77	0.91	0.72	0.94	0.88	0.96	0.83	0.94	0.84
29	0.90	0.73	0.90	0.68	0.94	0.87	0.96	0.83	0.94	0.82
30	0.87	0.70	0.85	0.72	0.97	0.93	0.92	0.79	0.99	0.96
31	0.93	0.67	0.87	0.72	0.97	0.87	0.96	0.79	0.96	0.88
32	0.93	0.63	0.79	0.49	0.97	0.84	0.96	0.88	0.94	0.87
33	0.90	0.70	0.76	0.51	0.91	0.82	1.00	0.96	0.93	0.83
34	0.80	0.57	0.78	0.54	0.91	0.82	0.96	0.79	0.97	0.88
35	0.80	0.60	0.78	0.54	0.88	0.78	0.96	0.67	0.87	0.74
36	0.87	0.63	0.74	0.56	0.88	0.78	0.88	0.79	0.93	0.82
37	0.77	0.67	0.57	0.38	0.84	0.68	0.79	0.63	0.84	0.71
38	0.73	0.43	0.50	0.40	0.65	0.47	0.58	0.46	0.72	0.52
39	0.70	0.50	0.49	0.26	0.46	0.28	0.54	0.46	0.65	0.43
40	0.27	0.17	0.18	0.13	0.44	0.29	0.42	0.29	0.53	0.34
41	0.27	0.20	0.15	0.12	0.22	0.15	0.29	0.21	0.35	0.19
42	0.20	0.67	0.10	0.04	0.18	0.09	0.25	0.08	0.29	0.16
43	0.03	0.03	0.06	0.04	0.12	0.06	0.04	0.00	0.16	0.12
44	0.10	0.07	0.10	0.03	0.12	0.09	0.08	0.08	0.21	0.10
45	0.17	0.10	0.09	0.04	0.10	0.06	0.08	0.04	0.06	0.04

According to Verma and Robertson (1973), the probability of all four months (June to September) receiving a rainfall above median is greater than the probability of them all being below median. Also, the probability of a day with 1.0 to 2.5 mm increases after June 10, while probability of a day with 5.0 to 10.0 mm rainfall increases after June 20 and continues up to September 20. Also, higher values of probabilities of wet day following a wet day during the monsoon shows the persistence in day to day weather, *i.e.* today's weather is not completely independent of yesterday's. The weekly rainfall totals (1931-1970) were analyzed for Markov chain model to know how much total rainfall in a particular week was expected, and the results obtained are shown in Table 11.

Table 11. Weekly total rainfall with 0.5 probability in different meteorological weeks

Met. Week	Expected with P=0.5	Rainfall (mm)	Met. Week	Expected with P=0.5	Rainfall (mm)	Met. Week	Expected with P=0.5	Rainfall (mm)
1	0	0 to 26	19	0	0 to 2	37	30	0-231
2	0	0 to 15	20	0	0 to 15	38	30	0-238
3	0	0 to 17	21	0	0 to 35	39	8	0-335
4	0	0 to 29	22	0	0 to 160	40	3	0-132
5	0	0 to 6	23	0.5	0 to 118	41	0	0-102
6	0	0 to 1	24	11	0 to 192	42	0	0-99
7	0	0 to 6	25	18	0 to 224	43	0	0-28
8	0	0 to 7	26	50	0 to 206	44	0	0-79
9	0	0 to 10	27	62	0 to 277	45	0	0-79
10	0	0 to 9	28	45	1 to 329	46	0	0-142
11	0	0 to 11	29	46	0 to 335	47	0	0-37
12	0	0 to 25	30	51	0 to 256	48	0	0-51
13	0	0 to 13	31	46	1 to 249	49	0	0-32
14	0	0 to 3	32	46	0 to 284	50	0	0-99
15	0	0 to 4	33	36	0 to 343	51	0	0-37
16	0	0 to 23	34	30	0 to 272	52	0	0-11
17	0	0 to 42	35	27	0 to 227			
18	0	0 to 11	36	30	0 to 252			

The data on values of the lowest and highest weekly total rainfall received in different meteorological weeks and that of rainfall expected with probability 0.5 show that any week during the year may be completely dry (without rain) and there is a big variation in the range of weekly total rainfall from year to year. Also, this range is much greater from 22nd meteorological week to 42nd week (last week of May to 3rd week of October) indicating that much less rainfall is expected before last week of May and after 3rd week of October. Interestingly, 1st to 3rd weeks of November and 2nd week of December receive more rainfall than other dry meteorological weeks of a year.

As regards the expected rainfall with probability 0.5, it is 0.8 mm for meteorological weeks (met. wks.) 1 to 22 (January to May end), 1.0 mm for met. wk. 23 (1st week of June), 11 mm for met. wk. 24 and 18 mm for met. wk. 25 (3rd week of June). The amount of expected rainfall increases to 50 mm for met. wk. 26, to 62 mm for met. wk. 27, but comes down to about 45 mm for met. wks. 28 to 32, and to about 30 mm for met. wks. 33 to 38, to 8 mm for met. wk. 39 and to only 3 mm for met. wk. 40 (1st week of October). Further no rainfall is expected from 41 to 52 met. wks. That is, normally there is no rain from 1st week of October to end of May.

3.2.8. Prediction of Rainfall

The rainfall data (1923-70) available at Indore were analyzed in different ways to see if it was possible to predict the rainfall in advance (Verma and Robertson, 1973):

3.2.8.1. Fitting Sin and Cos curves to rainfall data: To see if annual, 2-year, 5-year, 7-year, 11-year, 22-year, 23.5-year, 47-year, 94-year and 188-year cycles were followed by monthly rainfall data, Sin and Cos curves were fitted to these. Of various cycles, the annual cycle was the one showing the best fit. Thus, it was not possible to predict monthly rainfall.

3.2.8.2. Working out monthly and seasonal rainfall correlations: The correlation coefficients between the monthly rainfall for individual months (January to December) *vs.* total rainfall from June to November, July to November and August to November, were worked out to investigate the value of rainfall for a given month of the calendar year as a predictor of rainfall during subsequent months or seasons within the calendar year. No significant correlation was noted between any monthly rainfall *vs.* total rainfall unless the total rainfall included the rainfall of the predictor month. When the month in question (predictor or independent variable) is included in the predictant (or dependent variable) the resulting high correlation does not necessarily indicate that the month in question is a good predictor. Thus, the seasonal rainfall could not be predicted by the rainfall of any month.

3.2.8.3. Working out weekly and seasonal rainfall correlations: In order to see if rainfall of any beginning week was related to the seasonal rainfall, correlation coefficients were worked out between these two. The results showed the same trend as monthly rainfall correlations in general. But rainfall of first week in July (met. wk. 27) seemed to be indicative of seasonal rainfall to some extent.

3.2.9. River flows

Many important river systems flow through Central India. The biggest river system is Ganga and its tributaries flowing through Uttar Pradesh and Bihar. Its average annual flow is about 52 M ha m, of which about 50 percent (26 M ha m) is utilizable. Many important tributaries join it at right and left banks. The main tributaries joining at left bank are Gomati, Ghaghra (with its tributaries Sarda and Karnali), Gandak and Kosi. Theses tributary rivers drain not only Uttarakhand, Uttar Pradesh and Bihar states of India but also the Nepal country. The important tributaries joining at the right bank are Yamuna with its numerous tributaries and Sone river. Thus, Ganga basin forms the biggest part of Central India. It consists of Uttar Pradesh and Bihar states, north western and northern part of Madhya Pradesh and Chhattisgarh and north western and northern part of Jharkhand. Madhya Pradesh and Chhattisgarh states have the privilege of being associated with many major river systems of the country. Either these rivers or their tributaries originated from these states, or flowed through them after their origin from some neighboring state. While Mahananadi,

Tapti, Narmada, Mahi and Anas were major rivers originating from Madhya Pradesh, the major tributaries originating from the state were Tons, Sone and Rehand of Ganga; Chambal, Sindh, Betwa (with its tributaries Pahuj and Dhasan) and Ken of Yamuna, Pench, Wardha and Wainganga of Godavari; and Seonath, Hasdeo, Sender, Mand, Ib and Sonkh of Mahanadi. Chambal is very important tributary of river Yamuna originating near Mhow (Indore District of Madhya Pradesh), flowing through Rajasthan and forming boundary line between Rajasthan and Madhya Pradesh at first and then between Madhya Pradesh and Uttar Pradesh before joining Yamuna. Many important tributaries of Chambal originate in Madhya Pradesh and flow through Madhya Pradesh *viz.* rivers Chamla, Kshipra with its tributary Gambhir, Chhoti Kali Sindh, Kwari, Asan, Sank, Seep, Visali and Kuno, while others *viz.* Kali Sindh (with its tributaries Lakhandar and Parwan), and Parbati originate in Madhya Pradesh but flow through Rajasthan and join river Chambal in Rajasthan. Newaj is an important tributary of river Parwan originating in Madhya Pradesh but joining Parwan river in Rajasthan. Besides, some rivers originated from the neighboring state of Orissa but flowed through the state, for example, Indravati and Sabri, which are also the tributaries of Godavari river.

The next important basin is of Narmada river flowing through a major part of the region. The basin is long stretch of Narmada Valley between Vindhya ranges and Satpura hills. Average annual flow of Narmada river is 45.64 km^3 $year^{-1}$ and a major part (34.5 km^3 $year^{-1}$) is utilizable. Tawa and Barna are two important tributaries of Narmada. Tawa joins Narmada at Hoangabad. The south eastern part of Jharkhand forms the basin of Damodar river, while Central part of Chhattisgarh is part of Mahanandi basin. Mahanandi has an average annual flow of 67 km^3 $year^{-1}$, of which about 50 km^3 $year^{-1}$ is utilizable. Godavari basin covers (i) southern part of Chhattisgarh (Bastar plateau) drained by Indravati river – a tributary of Godavari river and (ii) Seoni district and part of Balaghat district of Madhya Pradesh drained by Wainganga – another tributary of Godavari river. Average annual flow of Godavari is about 110 km^3 $year^{-1}$of which 76 km^3 $year^{-1}$ is utilizable. As already said, Tapti river originates from Betul district of Madhya Pradesh and drains out very small part of Central India region south of Satpura hills.

As already stated, the monsoon rains are confined to June - September period normally and therefore maximum river-flows and surface run-off occur during this period. Also it varies considerably from year to year. Further, till recently most of the tributaries were perennial but now some of them dry up in summer. As regards the total quantity of surface water resource in the central states, only ad hoc assessments have been made from time to time. Madhya Pradesh Water Resources Development Committee headed by Shri N.D. Gulhati (1969) estimated the average water availability as 18.45 million ha m, whereas the Irrigation Commission (1972) reported the total flow of rivers in Madhya Pradesh and Chhattisgarh states to be 18.33 million ha m. If all this water was harnessed, which is almost impossible, it would irrigate about 9.5 million ha land (about 50 percent of the total cultivable area in the states) and also generate about 3000 megawatts (Mw) of cheap hydal power. According to Chaturvedi (1985), the estimated annual surface run-off from all major river basins was approximately 18.6 million ha m, and of this about 70 percent *i.e.* 13.02 million ha m could be harnessed for irrigation.

3.2.10. Groundwater potential and utilization

In Central India including both Madhya Pradesh and Chhattisgarh states, groundwater is the major source of fresh water for agriculture (irrigation), industry and even for domestic use. The quality of groundwater is, by and large, good except some specific spots where it is contaminated by industrial effluent, discharge of untreated waste water and municipal solid waste. The regime of groundwater in various regions of Madhya Pradesh and Chhattisgarh states needs to be studied over a number of years. Generally, most of the alluvial tracts which dominate in Central India, constitute the largest and most potential ground water reservoirs, as these contained water bearing strata in *kankar* (calcium carbonate nodules) layer, sand pebbles and gravel. Tube wells had been bored very successfully in the tracts. In Central Narmada valley, shallow tube wells commanding an area of 15 to 45 ha with an average of 25 ha, were successful. Ground water pockets also existed in trap as well as in rocky formation in rather confined state. Open dug wells could be constructed in this tract to irrigate about 3 ha only by one dug well. A very rough estimate of the ground water resources had been attempted by Central Groundwater Board (2006) and the data are presented in Table 12.

Table 12. Ground water status

S. No.	States	Annual Ground water recharge (billion m^3)	Net ground water availability (billion m^3)	Annual draft (billion m^3)	Draft as % of available
1	Madhya Pradesh	37.2	35.2	17.1	48
2	Chhattisgarh	14.9	13.7	2.8	20
3	Bihar	29.2	27.4	10.8	39
4	Jharkhand	5.6	5.3	1.1	21
5	Uttar Pradesh	76.4	70.2	48.8	70
	Total	163.3	151.9	80.6	53

As evident from Table 12, total annual ground water recharge from all the sources namely recharge by rainfall, canal, etc in Central India was 163.3 billion cubic meters and the net ground water availability 151.9 billion cubic meters. But the ground water development of the region was only 53 percent. It was very poor (just about 20 percent) in Chhattisgarh and Jharkhand states, and 39 percent in Bihar. In Madhya Pradesh also it was 48 percent *i.e.* less than 50 percent of the net available quantity. This 2006 assessment of ground water status by Central Ground Water Board differs from their earlier estimate according to which the total groundwater contribution from all the sources namely recharge by rainfall, canals, etc. was 55.1 billion m^3 for Madhya Pradesh and Chhattisgarh states, whereas the net groundwater recharge was 33.1 billion m^3

per year. Also, according to Chaturvedi (1985), the annual groundwater recoverable from recharge had been assessed at approximately 59.5 billion m^3 and it was estimated that about 50 per cent of the same could be harnessed for irrigation. In terms of area that could be irrigated, the ultimate groundwater potential of the states was estimated to be only 1.62 million ha by Irrigation Commission (1972) but as much as 3 million ha by Michael (1983). Apparently, there was some variation in figures on underground water resource estimates by different agencies. In spite of this variation in estimates, adequate quantity of underground was available in whole region (including both Madhya Pradesh and Chhattisgarh states). In Uttar Pradesh, a major part of the cultivated area (9.4 million ha) is irrigated through tube-wells and open dug wells. The hectarage of cultivated area irrigated in Chhattisgarh, Madhya Pradesh, Bihar and Jharkhand are 0.2, 3.7, 2.3 and 1.2 million ha, respectively. This shows that utilization of groundwater was very poor in Chhattisgarh and poor in Bihar and Jharkhand. As regards Madhya Pradesh a significant increase in exploitation of underground water through open dug wells and tube-wells (for domestic consumption as well as irrigation) and decrease in tank irrigation has resulted in depletion of underground water, particularly in Malwa and Gird Zone. Apparently there was an urgent need for augmenting the ground-water recharge through various techniques.

3.2.11. Progress on utilisation of surface flow

In Madhya Pradesh and Chhattisgarh states, the progress on utilization of surface flow for irrigation has been rather slow. At the time of formation of Madhya Pradesh, out of 156.41 lakh ha net sown area, only 8.29 lakh ha *i.e.* 5.9 percent of net sown area was irrigated. On the basis of gross cropped area and gross irrigated area, the percentage was only 4.9. After a decade, of the 165.42 lakh ha net sown area, only 9.76 lakh ha were irrigated *i.e.* 5.9 percent, and even after another decade only 9.6 percent of the net area sown in Madhya Pradesh was irrigated whereas of the gross cropped area, the gross irrigated area was only 8.9 percent. Up to 1984-85 *i.e.* 28 year after the formation of Madhya Pradesh state only 15.7 percent of the net area sown was irrigated. According to Bansil (2001) the irrigation potential created up to 1994-95 was 4.874 million ha (2.157 million ha under major and medium projects and 2.717 million under minor projects) but its utilization was only 3.976 million ha (1.529 million ha under major and medium projects and 2.447 million ha under minor projects). Even up to 2006-07, only 33.3 percent of the net area sown was irrigated by all sources (canals, tanks, wells, tube wells) and just a meager 5.7 percent through canals. In Chhattisgarh only 22.6 percent of net area sown was irrigated by all sources and as much as 19.0 percent through canals. The data on irrigated area in different central states through various sources of irrigation, *viz.* canal, tanks, tube-well and open dug well and others, were presented in Table 13. To compare the extent of different sources, the area irrigated through a particular source as per cent of total irrigated area in the state was worked out and shown within brackets, in the same table.

Table 13. Area (000ha) irrigated from different sources

State	Source of irrigation			
	Canal	Tanks	Tube-Wells & Wells	Others
Madhya Pradesh	976 (17.2)	127 (2.3)	3734 (66.0)	820 (14.5)
Chhattisgarh	768 (72.4)	50 (4.7)	165 (15.6)	77 (7.3)
Bihar	966 (27.9)	111 (3.2)	2251 (65.0)	134 (3.9)
Jharkhand	743 (31.2)	147 (6.0)	1165 (48.9)	328 (13.9)
Uttar Pradesh	2747 (22.2)	67 (0.5)	9371 (75.6)	206 (1.7)

As shown by the data in Table 13, canals were the major source of irrigation in Chhattisgarh state only (72.4 percent of the total irrigated area) where as tube-wells and open dug-wells were the main source in Uttar Pradesh, Madhya Pradesh and Bihar. The area under tank irrigation was miserably low in all the states. It was negligible in Uttar Pradesh (0.5 percent), and 2.3, 3.2, 4.7 and 6.0 percent in Madhya Pradesh Bihar, Chhattisgarh and Jharkhand (Sharda and Ojasvi, 2003).

This clearly shows that development of canal irrigation by using river-flows had taken place at snails speed in all the states of Central India, while development of irrigation through use of underground water by open dug wells and tube-wells had occurred by strides to the extent of excessive exploitation of underground water leading to enormous recession of water-tables at many places. A decrease in number of tanks and the consequent less area under tank-irrigation has aggravated this situation by fall in recharge of underground water. This points out the importance of run-off collection into tanks for irrigation as well as recharging the underground water.

3.3. CLIMATE

India has tropical monsoon climate with two monsoon systems namely south west monsoon and north east monsoon operating in it, during June to September and December to April, respectively. Central India also has tropical climate but only one monsoon system *i.e.* south west monsoon operates in most parts bringing rains from June to September usually. The north east monsoon operates only occasionally in some parts bringing winter rains during December to March.

Besides adequate rainfall, the region has three distinct seasons, *viz.* summer, winter and rainy season enjoying lot of sunshine and varying temperature regimes. The summer is hot and dry, winter cold and mostly dry while rainy season is mostly hot and humid. In summer, hot and desiccating fast wind shoots up the atmospheric maximum temperature to 48 to 50°C although the normal maximum temperature range is 35 to 45°C while freezing chilly winds in winter bring it down to 0 to -3°C from normal minimum temperature range of 10 to 15°C. Early hot wind (in March) could hasten the maturity of winter crops adversely affecting their yield, while low temperature below freezing point could result in frost damage to some crops.

3.4. CROP ZONES

With varied conditions of soil and climate, Central India represented favourable conditions for a wide variety of crops, and the same were grown in different parts of the region. The choice of traditional crops appeared to depend solely on the total amount of rainfall received in various parts of the region. For example, rice crop was concentrated in areas receiving a total rainfall higher than 120 cm, while wheat was confined to central flat lands with deep soils receiving an annual rainfall of 100 to 120 cm. The rice-wheat zone was a transitional zone between rice and wheat crop-zones. Sorghum-wheat zone in the northern and maize-wheat zone in the southern part of the region were situated in 80 to 100 cm rainfall belt while cotton - sorghum was concentrated in western Madhya Pradesh and Uttar Pradesh receiving an average annual rainfall less than 80cm. Based on area under different crops, the region was divided into seven main crop zones, *viz.* paddy zone, paddy - wheat zone, sorghum - wheat zone, pearlmillet- mustard, maize – wheat, wheat zone, and cotton - sorghum zone. Thus, the biggest emphasis had been on rice, wheat, sorghum, pearlmillet, mustard and cotton crops in the past but recently soybean has emerged as an important crop, particularly in Indore, Ujjain, Dewas, Sehore, Hoshangabad, Shajapur, Raisen, Bhopal, Betul, Chhindwara, Dhar, Rajgarh and Narsinghpur districts of Madhya Pradesh.

3.5. AGRO-CLIMATIC ZONES

On examining the topography, soil types and rainfall pattern of various crop zones, it became apparent that the whole of the area under a particular crop zone was not homogenous. Therefore, it might not be worthwhile to make the crop zone a basis for identifying and solving the agricultural problems; for example, the vast cotton–sorghum crop zone in Madhya Pradesh consisted of three distinct agro-climatic zones *viz.* (i) Malwa plateau, (ii) Nimar valley, and (iii) Jhabua hills. Thus, it was more appropriate to divide the whole region into fourteen homogenous agroclimatic zones for planning of agricultural development particularly of rainfed farming as shown in Fig. 2. Various agro-climatic zones are:

1. Western Gangetic Plain
2. Eastern Gangetic Plain
3. Mahanandi basin or Chhattisgarh Plains including Wainganga Valley
4. Bastar Plateau
5. Northern hill region (of Chhattisgarh) and Chhotanagpur Plateau
6. Kymore plateau and Satpura hills

7. Gird (Chambal valley) and Awadh zone
8. Bundelkhand region
9. Vindhyan plateau
10. Central Narmada valley
11. Satpura plateau
12. Malwa plateau
13. Nimar plains, and
14. Jhabua hills

Various agro-climatic regions are discussed briefly in continuation.

Fig. 2. Agroclimatic zones of Central India

3.5.1. Western Gangetic Plain

The zone covers north western and Central Uttar Pradesh having deep alluvial soil (Inceptisol) with loam to clay loam texture, susceptible to water erosion and neutral to alkaline in soil reaction. The northern part of the zone is sub-Himalayan region called Terai with high water table. Along major rivers occurs riverine land called *khadar* - low land covered by recent alluvium, where as land under old alluvium is upland called *bangar.* The climate is semi-arid, as the long term average rainfall of the region is 842mm (range 750 to 900 mm). Maize, sorghum, rice and *kharif* pulses are mostly grown in *kharif* while wheat, gram, and mustard are main *rabi* crops under rainfed conditions. By and large the soil is well-drained and productive but problems of underground water depletion and micronutrient deficiencies are coming up.

3.5.2. Eastern Gangetic Plain

The zone comprises of eastern Uttar Pradesh and Bihar and has deep alluvial soil (Inceptisol) with loam to clay loam texture. North western corner of Bihar (West Chamaparan) has eastern Himalayan foot hills and side slopes covering a small area of 30000 ha under well drained loam soils (Chandra, 2006). In Bihar, large areas are under *Tal* (10340 km^2) and *Diara* (11600 km^2) lands. *Tal* land has saucer shaped topography with no external drainage and therefore remains submerged under water throughout the rainy season. *Diara* land is undulating with no pattern of slopes. *Chaurs* land also occupies about 4000 km^2 area and remains under water for varying periods during rainy season. The climate is sub-humid with an average annual rainfall of 1050 mm. Rice, wheat, maize, gram and pea are main rainfed crops grown in the zone. Floods, waterlogging and improper drainage (external) are specific problems of the zone.

3.5.3. Chhattisgarh Plains including WaingangaValley

The vast Mahanadi basin is popularly known as Chhattisgarh plains covering Raipur, Bilaspur, Durg and Rajnandgaon districts and Raigarh, Sarangarh and Ghargoda subdivisions of Raigarh district and Kanker subdivision of Bastar district. It is a zone of high average annual rainfall ranging from 125 to 150 cm. The soils of the area are classified as mixed red and yellow including some skeletal soils on hill tops and slopy land. The soils are generally light colored and sandy on slopy land but heavy soil occurs on flat areas. The entire region is covered under paddy crop - zone.

The Waraseoni subdivision and *Garhi* region of Baihar subdivision both parts of Balaghat district, and *Kurai* subdivision of Seoni district constitute the Wainganga valley which stretches for several kilometers on either side of Wainganga, Beg, Deo and other rivers. The soil is medium to deep black (*kanhar*) formed by the alluvium brought down by these rivers from basalt hillocks. In the Wainganga valley, the soil resembled more closely the Chhattisgarh plain soils. This was an important rice tract of the Chhattisgarh state growing high quality paddy.

The principal crop grown on medium soil under bunded condition is rice, while on light and skeletal soils with good drainage, crops like minor millets and pigeonpea are extensively grown in *kharif* season. Under rainfed conditions, rice crop is followed by *utera* sowing of blackgram, greengram and linseed replacing lathyrus planted earlier, while under irrigated conditions wheat after rice is finding better prospects giving higher return per unit area.

3.5.4. Baster Plateau

Baster Plateau consists of entire district of Baster. The zone is characterized by hot sub-humid climate-hot summers, cool winters and heavy rainfall (100-160 cms) with annual mean of 156 cm. This zone is quite heterogeneous in topogr-aphy with undulating hilly tracts and flat valleys. Four distinct physiographic regions are:

i. Northern lowland upto Keshkal escarpment with a steeply rising relief from 150 m to 450 m above MSL.
ii. North eastern plateau extending from Abhujmad hills to eastern boun-dary of the zone with general relief from 500 m to 750 m above MSL.
iii. Southern plateau including Dantewada, Bijapur and Konta tehsils with relief varying from 300 to 600 m above MSL.
iv. Southern lowland stretching from Sukma, Konta to Bhopalpatnam as approximately 25 km wide strip with relief from 150 to 300 m above MSL.

The soils are very shallow and light textured red soils on ridges and plateaus and deep loamy lateritic and laterites in valleys. These are non-calcareous and slightly to moderately acidic, and have low CEC. Red soils have higher silica sesquioxides ratio than lateritic soils.

The principal crops of the zone are rice, minor millets in *kharif* and wheat, rapeseed, mustard and niger in *rabi* season.

3.5.5. Northern Hills of Chhattisgarh and Chhotanagpur Plateau

This zone marks the northeastern projection of the Indian Peninsula, and represents the major component of the eastern plateaus. It is akin to Satpura hills and extends beyond Maikal ranges to southwestern West Bengal althrough Jharkhand State. The zone covers Shahdol, Sarguja, Mandla districts, upper part (Jespur and Dharmajaigarh tehsils) of Raigarh, southeastern part (Singrauli tehsil) of Sidhi district, and most of Jharkhand State including Ranchi, Hazaribag, Singhbhum, Dhanbad, Palamau, Chatra, Giridih, Deoghar, Dumka, and Gumla districts and parts of Munger, Godda and Sahibganj districts (Sarkar, 2002). This is another big hilly region of central India, constituted of hill ranges, inter hill valleys, piedmonts, flood plains and inland valleys. The altitude ranges from 293 m to 1200 m above MSL.

Chhotanagpur plateau consists of a series of plateaus with different elevations. The highest part is Ranchi plateau including Pat region, covering whole of Ranchi district, and Southern part of Palamau district, and it is 600-

1100 m above MSL. The next is Hazaribag plateau (600-750 m above MSL) separated from Ranchi Plateau by Damodar trough. The Lower Chhotanagpur Plateau (300 m above MSL) covers outer parts of Chhotanagpur. The lowest and the non-uniform part consists of river valleys and plains 150-300 m above MSL.

The climate is hot sub humid having annual precipitation from 1050 to 1600 mm. The potential evapo-transpiration varies from 1400 to 1800 mm with an annual water deficit of 500 to 700 mm.

The soils are clay loam to sandy in texture and varying in colour from yellow to red on topo-sequence and categorized as Alfisol and Entisol in taxonomic nomenclature. The important soil-groups recognized locally in northern Chhattisgarh are: (i) highly eroded coarse textured (stony to sandy) soils covering about 10 percent of the cultivated area, (ii) Goda/Tikra soils being light textured and erodible have poor moisture retention and occupy 30 percent of cultivated area, (iii) Chawar: Heavy soils of lowland with gentle slope and bunded condition occupying about 50 percent of cultivated land, (iv) Bahra soils formed by deposition of transported alluvium in extreme lowlands with silty texture occupy 10 per cent of cultivated area. Red and yellow colours of these soils are due to coatings of iron oxides haematite and limonite, respectively, on soil particles. The soil reaction is neutral to acidic (pH 4.5 to 7.0) and silica-sesquioxides ratio high. The soils have a low fertility status in general. As regards crops, minor millets are commonly cultivated under rainfed conditions in *kharif* on 97 to 99 percent of cultivated area. Wheat and rice crops are grown under irrigated conditions in *rabi*.

3.5.6. Kymore Plateau Satpura Hills and Vindhyan Zone of Uttar Pradesh

This agro-climatic zone is quite heterogenous consisting of Kymore plateau, Satpura hills, and Vindhyan zone of Uttar Pradesh adjoining Rewa, Satna and Sidhi districts. The Kymore plateau lies between Panna and Kymore Hills and therefore consists of Rewa and Satna districts, most of Panna district, Gopadbanas and Deosar subdivisions of Sidhi district and Katni subdivision of Jabalpur district. Satpura Hills region includes the tracts of Satpura ranges in parts of Jabalpur and Mandla districts and in Seoni district. The altitude ranges from 300 to 900 m above MSL. The soils of this zone have been grouped into mixed red and black soils. In Mandla district, the shallow (skeletal) soils are locally called *barra*. Alluvial soil is found in the Vindhyan zone of Uttar Pradesh - the valley of Sone river. The climate is hot sub-humid (hot summers and mild winter) and the rainfall varies from 1250 to 1500 mm annually. Paddy and wheat are the major crops of the region. Presently, soybean in *kharif* and chickpea in *rabi* are grown on large areas.

3.5.7. Gird (Chambal Valley) and Awadh zone

Gird and Awadh zone falls under central highlands, and it is traversed by Vindhyan range hills. Chambal valley along the northern border of Madhya Pradesh in continuation of the Gangetic plain in south-west Uttar Pradesh constitutes a very important agricultural region. It consists of Agra, Mathura,

Aligarh, Hathras, Etawah districts of Uttar Pradesh, Gwalior, Morena, Sheopur and Bhind districts and parts of Shivpuri (Shivpuri, Kolaras and Pohri sub-divisions), Guna and Ashoknagar districts of Madhya Pradesh. The climate of the zone is characterized by hot and dry summer and cool winters. The rainfall ranges from 600 to 800 mm annually increasing from west to east. The region is subjected to heavy erosion resulting in the formation of ravines. Under rainfed conditions, sorghum and pearlmillet are sown mixed with red gram while chickpea, rapeseed and mustard, wheat and barley are sown in *rabi* after *kharif* fallow. Recently soybean crop has become popular on black soil of Guna and Shivpuri. However, 31.4 percent area of the zone is gross irrigated and is intensively cropped.

3.5.8. Bundelkhand zone

The zone covers seven districts of Uttar Pradesh (Banda, Chitrakoot, Hamirpur, Jalaun, Jhansi, Lalitpur and Mahoba) and three whole districts and parts of three districts of Madhya Pradesh *viz.* Datia, Tikamgarh and Chhatarpur districts, Karera and Pichor sub-divisions of Shivpuri district, Ishagarh and Chanderi blocks of Guna district and Saleha sub-division of Panna district. This zone is intersected by low hillocks and its elevation varies from 266 to 560 m above MSL. Because of undulated and rugged topography, the zone is prone to high degree of degradation. The occurrence of various soil types in the zone is closely related to its physiographic regions, *viz.* (i) hillocks (ii) plateaus (iii) piedmont and piedmont plains, (iv) alluvial flood plains and (v) gullied land. In general, soils seem to follow a topo-sequence of yellow – red – brown - black soils on hillocks, hill slopes plateaus and low lying areas. The zone has mostly mixed red and black soils. Red soils fall into orders Inceptisol, Entisol and Alfisol. The colour darkens depending upon the amount of black soil mixed with it. Red soils are further grouped into *rakar* and *parwa* soils. *Rakar* occurs on elevated spots at the foot of hills and is severely eroded, shallow, coarse textured and gravelly (*murrum*) poor in moisture retention. The soil reaction is slightly acidic to neutral. Parwa soil is brown to greyish brown, has pH 7.0 to 7.5 and a *kankar* ($CaCO_3$ nodules) layer below sub-soil. It is medium to deep loam soil. Black soils are Inceptisol and Vertisol taxonomically and are further divided into *Kabar* and *Mar* soils. *Kabar* is coarse textured and free from lime concretions. *Mar* is fine textured, very adhesive when wet and forms wide and deep cracks on drying. It has $CaCO_3$ layer in subsoil and is slightly alkaline (Verma et al 1996). Only 32.1 percent of the total cultivated area was irrigated, mostly by wells and tanks. It receives an annual rainfall of 800 to 1150 mm with a mean of 940 mm. The Potential Evaporation varies from 1400 to 1700 mm. Principal crops grown under rainfed conditions are sorghum, sesame, groundnut and soybean in *kharif* and gram, wheat and linseed in *rabi.* Because of highly variable rainfall, not only crops are adversely affected by drought but the groundwater also becomes scarce and the food security is endangered. Short rainy season with a few intense rains and extremes of temperature (-1.5 to 49°C) further add to misery. Bundelkhand is one of the most underdeveloped region of the country. The yield potential is low and agriculture is not viable unless appropriate measures of land and water management are adopted (NBSSLUP 1981).

3.5.9. Vindhya Plateau

Vindhya plateau zone is constituted of district of Damoh, Sagar, Vidisha, Bhopal, Raisen (except Bareli sub-division) and Sehore (except Budni sub-division) and part of Guna district (Chachoda and Raghogarh sub-divisions). Average elevation of the plateau is 500 m above mean sea level. The soil characteristics of the zone resemble those of Hoshangabad district (deep black soil known as *mariar* locally). The soil of Guna district falls under the broad soil group medium black soil of trap and gneissic origin and are generally loam to clay in texture. Overall, the soils are rated as deep black with less than 0.5 percent slope. The climate of the region is hot sub-humid characterized by hot summer and mild winter. The annual rainfall varies from 1000 to 1250 mm. Rainfed farming dominates in the region as only 16.2 percent of cultivated land is irrigated. The principal crops under rainfed conditions in this region are wheat, chickpea, lentil and linseed in *rabi* grown after *kharif* fallow and sorghum, maize, greengram and pigeonpea in *kharif*. With the introduction of soybean crop, *kharif* fallow has been reduced slightly. However, fallow - wheat is still quite common in the zone and consequently this is the wheat bowl of Madhya Pradesh, though mainly rainfed. Lot of area (up to 80 percent) in this agro-climatic region was left fallow in *kharif* and only one crop, mainly wheat was grown in *rabi* on profile stored moisture.

3.5.10. Central Narmada Valley

A narrow and long strip of land extending to 300 km in length and 80 km in width along Narmada river between Jabalpur on the east and Harda on the west consists of Narsingpur and Hoshangabad districts (except Harda sub-division), part of Jabalpur district, Budni sub-division of Sehore district and Bareli sub-division of Raisen district. The zone is marked by low plains of Narmada river and Satpura hills. Narmada valley has very deep black soils (Vertisols) with enormous shrinking and swelling properties. The soils of the entire region are formed by deep alluvial deposits. The topography is flat with even surface in its centre. Also there are patchy layers of archaean rocks represented by gneissic granites, phyllites etc. In Jabalpur, nearly 31 percent is occupied by good soils which include *kabar* and *mund* deep soils. Medium soils *i.e. domatta* and *sehra*, occupy 36 percent while light soils called locally as *pataura*, *bhata* and *barra* cover 32 percent. The climate of the area is hot sub-humid with hot summers and mild winters. The mean annual rainfall varies from 1000 to 1500 mm. Annual potential evapo-transpiration ranges between 1300 and 1500 mm. *Kabar* and *mund* soils (locally called deep black soils) are suitable for *rabi* crops especially wheat, while *mund* and *sehra* (local name for medium black soil) for rice. *Pataura* is light textured shallow black soil used for growing inferior crops like sorghum while sandy and gravelly *bhata* and *barra* (on hill slopes and piedmonts) are used for minor millets like kodon and oilseed like niger. The Narsinghpur soils are deep black and further sub-divided into black, dark-brown, coarse brown and red mixed sandy soils. The haveli system of cultivation prevails over large areas in the zone. The bunded fields were submerged under water in rainy season and wheat was sown after draining out water without much preparatory tillage.

3.5.11. Satpura plateau

The zone consists of the hilly region formed by Satpura ranges as well as the flat cultivated areas of Betul and Chhindwara district. Satpura range runs from west to east across the zone. The zone has an elevation of 600 to 1000 metres above mean sea level. The important rivers of the zone are Pench, Kanhan, Tapti and Machna. The zone has comparatively low temperature and annual rainfall in the zone varies from 1000 to 1250 mm. The soil of the zone is characterized by shallow and medium black soils of granite origin. The soils have clay percentage ranging from 15 to 30, and are called locally as *kali morand*, *dorsa*, *matasi* and *sehar kachhar*, etc. The irrigated area in the zone is 18.1 percent mainly through wells. It falls under maize-wheat crop zone, and maize, sorghum, pigeonpea and groundnut are the principal *kharif* crops raised on relatively less eroded and consequently more fertile soils, while minor millets were grown on comparatively poor and eroded soils. In *rabi*, wheat, gram and linseed are the principal crops. About 20 percent of the area is under oilseed crops, of which soybean is a major crop followed by groundnut.

3.5.12. Malwa Plateau

Malwa plateau is vast agro-climatic region covering the districts of Ratlam, Mandsaur, Indore, Dewas (except Kantlod and Kannod blocks), Ujjain, Shajapur, Rajgarh and Sehore and Badnawar and Sardarpur sub-divisions of Dhar district and Petlawad sub-division of Jhabua district. The region is bound by Vindhyan ranges on south and east sides, striking north from Bhopal to Chanderi and from Amjhere to Chittor on the west. Mukemdwara range strikes east from Chittor to Chanderi in north. The main rivers of the plateau are Chambal and its tributaries *viz*. Kalisindh, Shivna, Gambhir, Kshipra, Ajnar, Dudhi, Nevaj and Chhapi. The region has black clay soils usually loam to clay in texture with varying clay content from 25 to 55 percent. The clay soils are characterized by high moisture retention, free calcium carbonate (lime) concretions present at different depths. Cracks develop under moisture stress conditions, and this property is characteristic of the montmorillonite type clay mineral being dominant in these soils. The climate of the region is semi-arid with hot summers and dry winters. The average rainfall in the region ranges from 750 to 1250 mm. These soils are suitable for most of the crops, although the region fell in sorghum - cotton zone. The irrigated area in the region is 22 percent of the total cultivated area. Under rainfed conditions, most of the area was single cropped and wheat and gram crops were generally sown on *kharif* fallow in alternate years. Other important crops were sorghum, groundnut and maize. Recently, with introduction of soybean crop the area under *kharif* fallow and cotton crop has got reduced considerably and that under soybean increased.

Thus the zone is now sorghum-soybean crop zone and not cotton-sorghum zone.

3.5.13. Nimar Plains

Nimar plains situated between the Vindhya and Satpura range, west of Harda sub-division of Hoshangabad district constitute this zone commonly known as Nimar. It consists of district of Khandwa and Khargone, Harda sub-division of Hoshangabad district and Manawar sub-division of Dhar district. The major part of the zone lies between valleys of Narmada and Tapti rivers flowing parallel to each other from east to west. Nimar plains represent the south west part of Narmada valley which extends from Jabalpur to Barwani in Dhar district

Medium and deep black soils are mostly found in the zone. These are loamy to clay in texture. The climate of the region is characterized by hot summers and dry winter. The annual precipitation ranges from 500 to 1000 mm. The percentage of gross irrigated area is 21.7 percent. There are different local classes of soils of which *ranker* is the lightest soil having parent rock at a depth varying from 8 to 60 cm. It contains large proportions of pebbles, limestone and sand. The area was traditionally included in sorghum-cotton zone but presently groundnut occupies the largest area next to sorghum. Wheat is grown in *rabi* under irrigated conditions and its area is more than that of cotton. Along the river banks, well drained deposits are found which are quite susceptible to erosion (Verma, 1986).

3.5.14. Jhabua Hills

This is a peculiar hill zone constituted of Alirajpur, Jhabua, Jobat and Thandla sub-divisions of Jhabua district and Kuskhi sub-division of Dhar district. The region has medium black soils but it is characterized by undulating topography having slope from 1 to 50 percent. In general, 20 to 25 percent of total cultivated land has 2 to 3 percent slope, 30 to 35 percent of land has 3 to 20 percent slope and 40 to 45 percent of land has more than 20 percent slope. Nearly 65% area has very shallow soils on upland and slopes which are prone to erosion. There are 3 main rivers in the region, *viz.* Narmada, Mahi and Anas and each river has its own catchments. Mahi catchment in northern part of the zone covers an area of about 2000ha. The Anas river originates from a village in Jhabua subdivision and flows through Thandla sub-division to enter Gujarat state. The river has a catchment of about 7800 ha covering the central part of Jhabua district. The Narmada river demarcates the southern boundary of the district and has a catchment area of about 500 ha.

The climate of the zone is hot semi-arid with hot summers and dry winters. The rainfall varies from 500 to 1000 mm. Frequently dry spells during crop growing period make the region drought-prone. The gross area irrigated is 13.2 percent. The region is suited for growing maize, cotton, groundnut and millets although paddy is also grown in valleys.

IV

WATERSHED BASED RAINFED FARMING DEVELOPMENT

Agriculture of Madhya Pradesh, Uttar Pradesh, Bihar and Jharkhand as stated earlier, was dominantly rainfed, because only 17.3, 22.1, 27.9 and 31.21 percent (2006-07), respectively, of the net irrigated area in these states is canal irrigated. That is, most of the irrigation in these states depends on rain-water or underground water. Thus, the development of agriculture on more than 90 percent of the area in the states virtually depends on improvement of its rainfed farming. Realizing the importance of this, the State Governments and the Indian Council of Agricultural Research have financed research projects on dry farming which are still operating at Agra, Faizabad, Jhansi, Varanasi, Jagdalpur, Ranchi, Indore and Rewa since 1970. The Government of India financed pilot projects on rainfed farming development, while the Indian Council of Agricultural Research in collaboration with U.K. Government sponsored Indo-U.K. Dry Farming Operational Research Project (ORP) on Black soils to transfer and test the technology evolved by research to farmer's fields. Subsequently, other ORPs were also launched. Several important aspects of the rainfed farming technology such as land and water management practices, cropping patterns, fertilizer use and plant protection measures including seed treatment were tried under farmers' own conditions. Farm machinery and bullock-drawn implements evolved in U.K. and our country, respectively, were also tested on farmers' fields as part of the project programme. Thus, valuable information derived by research work on small plots at various research centres had been tested and practiced on the farmers' fields. The research results and the experience gained on farmers' field had been synthesized into a technology which was suitable to farmers for the development of mixed farming under rainfed conditions.

4.1. RATIONAL APPROACH FOR DEVELOPMENT OF RAINFED FARMING

Rainfed farming meant growing of crops using the rainwater as the sole source, *i.e.* the crop-growth was completely dependent on rainfall and the profile-stored moisture, and even the supplementary irrigation was not used. Although strictly under rainfed conditions also, there was a big scope for increasing crop-production through cropping in *kharif* (when moisture was not a limiting factor usually) and adopting a package of improved practices, a revolution in rainfed farming can be brought about by runoff collection and recycling the rain-water in addition to other improved practices. Under rainfed farming, the cropping pattern and the crop-yields were determined by moisture availability period which in turn was determined by total rainfall, intensity of rainfall, distribution of rainfall (number of rainy days), land slope, soil type (particularly its infiltration rate and water holding capacity), depth of soil and finally the land water-management practices adopted by the farmers. The crops and their varieties grown by the farmers were important to make the most efficient use of the available water. Since rainfall was the only source of water for crop-growth under rainfed farming, its efficient use was the prime factor deciding the success and failure of crops under rainfed conditions. Where the rainfall was sufficient and well distributed throughout the year, the rainfed farming was very successful, *i.e.* the crop yields were not only high but were stable also. But in those regions where the rainfall was erratic in distribution over time and space, or evapo-transpiration was higher than precipitation, the problem of rainfed farming was really difficult to tackle. Under such conditions, the success of rainfed farming depended upon how efficient was the conservation of rain-water and soil moisture and its utilization for crop-production. Thus, essentially the strategy was to manage "rain-water resource" considered as part of the whole resource management system. Under rainfed condition, rain-water resource can be best managed on watershed basis, because rain-water starts flowing from where watershed begins.

4.2. WHAT IS A WATERSHED?

Before discussing watershed development, it is necessary to know what a watershed is. A watershed (also called a catchment) is the entire land area from which surface water drains into a stream at a particular point. Watershed is basically an independent geo-hydrological unit encircled by a boundary of ridge line upto a point in valley of a stream. Watershed can be of any size depending upon whether it is of (i) a main river, (ii) its tributary, (iii) a small stream or (iv) a *nala*. Thus, the size of a watershed varies from a few hectares to millions of hectares depending upon the location on a stream at which the watershed is to be measured, and the watershed of a river consists of numerous mini watersheds of various tributaries and a very large number of micro-watersheds of different *nalas* joining the rivulets (Fig. 3).

4.3. WHY WATERSHED APPROACH?

A watershed is the entire area which contributes runoff water to a given location on a stream and therefore, to manage this water, a watershed is the most appropriate unit. The runoff water can be fully quantified for the watershed and taken care of. Also, the management of rain-water has to be thought of right at the point it starts flowing, and this is where watershed originates. Thus, the management practices can be adopted starting from the ridge and extended to area downwards in the direction of water flow, with time. At first, one micro-watershed, then other micro-watershed of a mini-watershed, one by one, can be selected for development. Thus, watershed approach is very appropriate for the development of rainfed farming in which the land is developed micro-watershed by micro-watershed for making the most efficient use of rain-water.

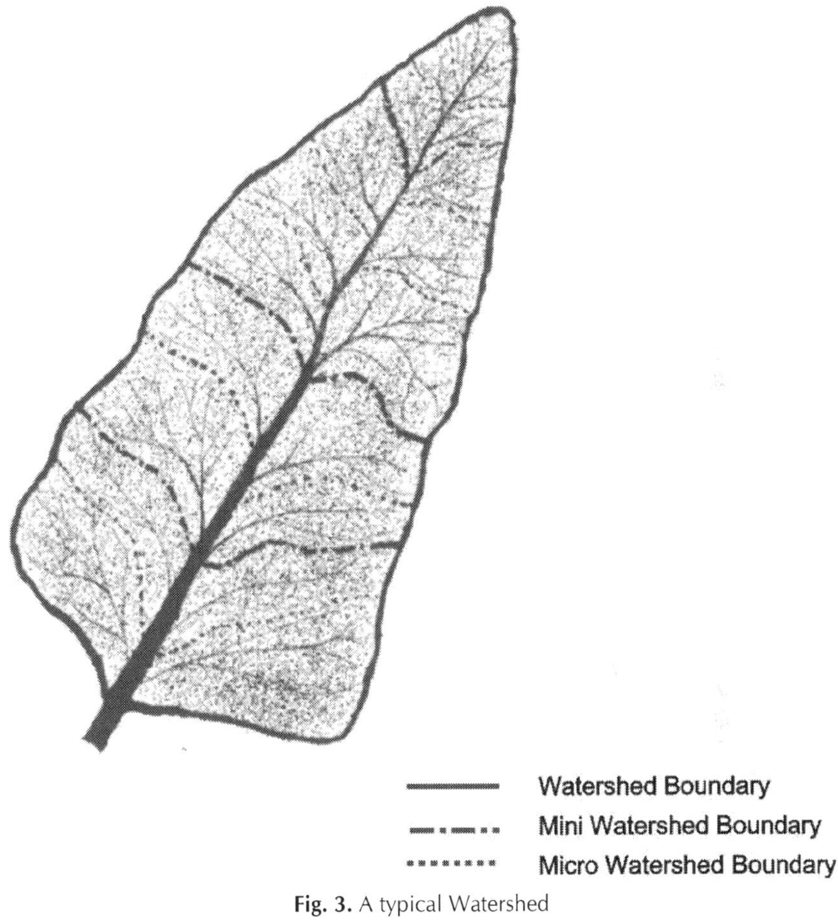

Fig. 3. A typical Watershed

4.4. CLASSIFICATION OF WATERSHEDS

According to size of watersheds, these could be classified into:

1. Macro-watershed	:	greater than 50,000 ha
2. Sub-watershed	:	10,000 to 50,000 ha
3. Milli-watershed	:	1000 to 10,000 ha
4. Mini-watershed	:	100 to 1000 ha
5. Micro-watershed	:	less than 100 ha

4.5. WHAT IS WATERSHED MANAGEMENT?

Watershed management means management of resources such as land, soil and rain-water on a watershed basis. The objectives of watershed management are:

1. To manage land and water resources of the watershed for their optimum utilization.
2. To manage land to manage rain-water to control the problems of soil erosion, inadequate drainage, maintenance of fertility and conservation of water.
3. To obtain optimum upland treatment (conservation practices) consistent with efficient land use and high income per unit area.
4. To make a substantial contribution towards flood prevention.

4.6. WATERSHED PLANNING

Appropriate management of land and water resources on watershed basis requires a scientific planning so as to prepare a master-plan. Various steps to watershed planning are:

1. Identification and demarcation of watershed and divide it into mini/micro watersheds.
2. Collection of data on basic resources (resource inventory).
3. Identification of problems of the micro/mini watersheds.
4. Assessment of available technology to solve the problems identified.
5. Preparation of master-plan.

4.6.1. Identification and demarcation of watershed

For preparing a master-plan for the development of a watershed, it was necessary to identify the watershed and demarcate its boundaries. Once a particular micro-watershed or a mini watershed was selected for development, its boundary could be identified by first marking the main drainage line and then identifying the ridges which formed the boundary of the main watershed. Subsequently the secondary and tertiary drainage lines could also be marked to divide the watershed into sub-watersheds and further sub-divide the sub-watershed into milliwatersheds/miniwatersheds and finally into micro-watersheds (Fig. 3). One of the following four methods could be used to mark the drainage lines:

a. Topo-sheet: Enlargement of the latest relevant topo-sheet of the biggest scale available (1:50,000) and then marking the smaller streams (not shown in the topo-sheets) by traversing the area and find the ground truth.
b. Aerial photographs: Stereoscopic pairs of photographs were available with Surveyor General of India and the same for a particular area could be obtained from him. By analyzing these photographs with a stereoscope, the drainage lines could be marked on the mosaic of aerial

photographs. The map thus obtained could be blown to any suitable scale for further use.

c. Topographical or Contour survey: A detailed survey of the site and area under consideration for catchment development must be made for mapping the contours of the area. The topographical survey had to be conducted at 30m grid or at still lower grid at special points, *viz.* gullies, very undulating grounds and hillocks. Contour map of area showing contours at 0.3 m or 0.5 m interval and other permanent features on the land such as well, trees, electric poles, house, temples, etc. should be prepared. Such a map could give a clear picture of the watershed, mini-watershed and micro-watersheds as the line joining the apices of convex curves represented the ridge while that joining the apices of concave curves represented the drainage line.

d. Village maps available with department of land records: Fortunately village maps (scale 1:400) are available for all villages. These can be reduced to a suitable scale (1:4000 or so) and a map of whole watershed prepared by joining together the maps of all villages in a watershed. This map will have all the drainage lines of the watershed. A map of scale 1:4000 for a watershed of about 3000 ha has been found very suitable for field-work.

e. Landsat imagery: Indian Space Research Organization (ISRO) has launched a variety of satellites to gather information/data on terrain, geomorphology, soil erosion, etc. of earth surface. The landsat imageries are available for various parts of the country. The relevant imagery can be analyzed to determine drainage lines of a watershed. For example, high resolution IRS-ICPAN data fused to LISS-3 data generated thematic maps showing land slope, drainage lines, land- use etc.

4.6.2. Resource inventory

Inventory of resources in a watershed was necessary so that plan could be drawn for their appropriate utilization. Collection of data on land, soil, rainfall, underground water, human and bullock power, etc. was necessary. For this different types of surveys had to be conducted, as described in continuation.

4.6.2.1. ***Soil Survey:*** Soil survey of the area was necessary for the preparation of soil map giving the information on distribution of soil series, soil types, soil depth, slope and degree of erosion. The soil survey report contained information on soil association named and defined in terms of taxonomic units, as well as on fertility status and erodibility of the soils in the watershed. Since the technique of land management in the watershed will solely depend upon the slope of land, it was necessary to demarcate land-slope classes *viz.* land with less than 1 percent slope, 1 to 3 percent slope, 3 to 6 percent slope and greater than 6 percent slope land. That is, the survey will divide the whole watershed into classes such as flat land (less than 1 percent slope) moderately sloping land (1 to 3 percent slope), highly sloping land (3 to 6 percent slope) and steep land (greater than 6 percent slope). Finally, the report contained information and map of various land capability classes in the watershed.

4.6.2.2. Underground water survey: To ensure stability of crop production and to increase crop yields, it was necessary to provide life saving irrigation to *kharif* crops and supplementary irrigation to *rabi* crops. Therefore, underground water was very valuable for rainfed areas, and it was necessary to have an estimate of the underground water available in the watershed, to prepare a plan for its rational utilization. Such information might be available with Central Groundwater Board also.

4.6.2.3. Meteorological data: Rainfed farming was dependent on rain-water and therefore, it was necessary to collect data on various weather parameters *viz.* daily rainfall, minimum maximum temperature, potential evapo-transpiration, relative humidity, rainfall intensity and rate of runoff. Since rain-water was the only source of water for rained crops, it was necessary to study the rainfall characteristics of the watershed. The data on daily total rainfall as well as on intensity of rainfall was necessary to derive different types of information by analyzing it, *e.g.* variation of rainfall in time and space, probability of different amounts of weekly total rainfall, amount of weekly total rainfall expected with 80 percent probability, excess and deficit of water during different meteorological weeks, and finally the calculation of moisture availability period in a watershed, so that crop-plan could be tailored to this period.

4.6.3. Socio-economic survey

Socio-economic survey is aimed at determining the farmer's resources such as human labour, farm-power, farm implements, livestock, investment capacity, financial condition (extent of indebtedness), source of financing and various agricultural and rural development agencies including input supplying ones operating in the watershed. Besides, complete information on present farm income, prevalent cropping pattern and crop yields, productivity of milch animals, income from other sources, extent of present employment, etc. was required to be collected. This information will serve as bench mark survey to spell out the socio-economic status of the watershed before development plan was executed.

4.6.4. Identification of problems of watershed

The problems of soil erosion, water-logging, water stagnation, low productivity of various crops and of the livestock with the farmers in the watershed must be identified. A detailed analysis of the problems and causes thereof had to be made for finding out the solution to these problems.

4.6.5. Assessment of available technology

After the problems of the watershed were identified, relevant technology evolved by research had to be assessed and the strategy to solve these problems had to be made. Technology in respect of land and water management practices, crop management practices and animal husbandry practices had to be applied according to problems existing in the watershed. If local experience of research was lacking, the results obtained under similar agro-climatic conditions elsewhere were to be considered for application as such or after adaptive trials on a smaller area in the watershed to start with and then spread to larger areas.

4.7. PREPARATION OF WATERSHED DEVELOPMENT PLAN

After all the information under paragraphs (a) to (f) was obtained and different maps incorporating the relevant information were prepared, a master-plan for the integrated development of watershed could be prepared (Verma, 1990). The master-plan will include:

a. Plan for land and water management indicating on a map the locations of approach roads (on ridges), storm drains (water diversion bunds), graded bunds (from crest to waterway), grassed waterways, surface drains, drop structures wherever necessary, structures for safely letting in run-off water diverted along graded bunds into waterways e.g. ramp, ramp-cum-waste-weir, grassed outlet, pipe outlet, etc. Also, sites for gully reclamation structures, structures for stabilization of natural streams, and run-off collection structures (submersible check dams, farm ponds, tanks etc.) must be shown at appropriate places.

b. Plan for inter bunded area treatment, if felt necessary to provide crop drainage (contour furrows, raised beds and sunken beds etc.) depending upon the land-slope, texture and depth of the soil, annual rainfall in the area, and choice of the crop to be grown;

c. Plan for judicious exploitation of underground water (through open dug wells or tube-wells depending upon the feasibility), and for use of run-off collected water;

d. Plan for appropriate land-use according to the suitability of the land: afforestation of bare hillocks and slopy land having very shallow soils with trees for fuel, fiber, fodder, fruits and timber; planting of improved grasses and legumes and legume shrubs on slopy land with shallow soil; planting of short duration crops/varieties on less slopy land with shallow soil; intercropping system on medium deep soils and appropriate crop-sequences on deep soils).

e. Plan for livestock improvement including growing of fodder crops, improving pastures, improving animal housing and then upgradation of scrub cattle through cross-breeding programme using artificial insemination or by direct service of a pedigree bull.

f. Plan for contacts: Before implementation of watershed development plan, complete rapport with farmers on whose fields the plan was to be implemented was necessary. This required mass contact with all the participating farmers in the form of village meetings as well as individual contacts, particularly with those farmers whose fields were located at the start (upper periphery of the watershed i.e. those farmers on whose fields the development work was to be done first of all. It was necessary to win the confidence of all the participating farmers for the successful implementation of the plan. Mass contacts with people of the watershed are necessary for (i) making the people aware about the watershed development programme to be executed on their fields, (ii) to discuss their suggestions and incorporate these into plan of action, (iii) to make sure that all the people are actively involved in the programme as there must be some item in the plan to benefit small and marginal farmers as well as the landless inhabitants of the watershed, (iv) to encourage people to form self help groups, and (v) to elect their representatives to "Programme Implementation Committee" at Watershed Level (Verma 1987).

g. Plan to contact public representatives of the area: Meeting with local M.L.A, M.P., Sarpanch of village Panchyat, Gram Sabha members, etc. to make them aware of the programme,

h. Screening of films on success stories of Watershed Development Projects executed elsewhere.

4.8. IMPLEMENTATION OF WATERSHED DEVELOPMENT PROJECTS

Although soil conservation research in India stared in 1949-50 with the establishment of a Soil Conservation Department under Damodar Valley Corporation, Hazaribagh, Bihar, watershed management came into focus much later-as late as early nineteen hundred seventies. Watershed based development of dryfarming/rainfed farming was really conceived in 1973-74 when the first Operational Research Project entitled Indo-British Operational Research Project (Indo-UK Dry Farming Project) was launched at Jawaharlal Nehru Krishi Vishwa Vidyalaya Campus College of Agriculture, Indore, Madhya Pradesh The Project had British Govt. collaboration and was funded jointly by the Indian Council of Agriculture Research and U.K. Govt. The land and water management programme of the project was implemented under the guidance of British Team Leader Prof. Norman W. Hudson (renowned Soil Conservation Engineer) at first and then Prof. T.F. Shaxson - Land Husbandry Expert. Eye opening results of integrated watershed management work undertaken by the Project prompted I.C.A.R to start three more ORPs at Fakot (Uttarakhand), Sukhomajari (Haryana) and G R Halli (Karnataka). Encouraged by grand success of ORPs, 47 model watershed based development projects were implemented by Central Research Institute for Dryland Agriculture (CRIDA) Hyderabad, in different agro-climatic regions of the country in 1983. Since then watershed based approach became the rule for Drought Prone Area Programme, Rainfed Areas Development Programme, Hill Area Development, Integrated Wasteland Development Project etc. Not only Ministries of Agriculture, Rural Development and Environment and Forestry, Govt. of India, but also International Agencies like World Bank, European Economic Commission (EEC) now named European Union (EU), Danish International Development Agency

(DANIDA), Japanese International Cooperation Agency (JICA), Indo - German and Indo - Swiss Agencies funded watershed development projects were executed. Further various state governments and NGOs also implemented watershed development programmes. Case studies of some of these projects are summarized below to bring out the fact that watershed based rainfed farming development technology evolved by research has paid great dividends.

4.9. CASE STUDY OF OPERATIONAL RESEARCH PROJECT, INDORE

The project was implemented in 3 phases:

i. Indo - British Operational Research Project (1974-79): The watershed developed was 2700 ha covering 3 villages viz. Jamburdi-Hapsi, Nainod and Rinjlai (Indo-UK DFP 1979, 1980b).

ii. Operational Research Project (ICAR) Indore, M.P. (1979-1986) extended by ICAR to cover lower parts of both sub-watersheds (J-Hapsi sub-watershed and Nainod-Rinjlai sub-watershed) in Hingonia and Pipliyatapha villages. As a matter of fact, the grand success of Indo-UK Dry farming project prompted ICAR to start Operational Research Projects at eight locations in the country including extension to Indore ORP in which the extended area was 424 ha towards Gambhir River – a tributary of Chambal river. The area had comparatively flat topography (slope 1 per cent and less) and deep black soil (Vertisol) - soil series Sarol.

iii. Pipliya Tapha watershed development project (1986-87 to 1995-96) ORP, Indore, was further continued to cover the remaining part of the watershed of Gambhir river about 400 ha (Tomar 2009). Achievements of the Operational Research Project, Indore, executed on a watershed of about 3500 ha in 3 phases were presented below:

4.9.1. Works executed

The following land and water management works were executed in the watershed:

i.	Water Diversion Bunds/Storm Drains	6 Kms.
ii.	Grassed water ways	10kms.
iii.	Gabion structures	22
iv.	Waste-weirs	40
v.	Loose Boulder structures	20
vi.	Run off collection Tanks	02
	Water storage capacity	(i) 10,000m^3
		(ii) 17,000 m^3
vii.	Graded bunds (ridge to waterways)	15,00 m
viii.	Spurs (Gabion walls)	2

4.9.2. Cost Estimates

Cost estimates (1979 price basis) for various land and water management works executed in the watershed were as follows:

a. **Storm drain:** Cost of storm drain with average cross section 1 m^2 – Rs. 10/- per running metre.
b. **Graded bund:** Cost of graded bund with average cross section of 0.5 m^2 – Rs. 5/- per running metre.
c. **Grassed water-way:** (i) Cost of construction of two bunds with average cross section 1.7 m^2- Rs. 34/- per running metre. (ii) Cost of planting grass on both bunds – Rs. 0.80 per running metre.
d. **Loose boulder structure** – Rs. 700 to Rs. 1000 each.
e. **Gabion structure** – Rs. 3000 to Rs. 20,000 depending upon the size.
f. **Cost of waste weir** – Rs. 300 to Rs. 500 each.
g. **Cost of run-off collection**– Rs. 4000 to Rs. 5000 per ha irrigated area.
h. **Overall cost-** Rs. 1500/- per ha (for watershed of 2000 ha covered under Phase 1).

4.9.3. Benefits

Appropriate land and water management and adoption of other improved practices resulted in not only increased agricultural production but also in lessening the fluctuations in crop yields due to aberrant weather conditions (Verma, 1982). The perceptible changes were as follows:

a. ***Kharif* fallow**: *Kharif fallow* reduced from 66.2 percent in 1974-75 to 32.2 percent in 1981-82.
b. **Intensity of cropping**: Cropping intensity increased from 108 percent in 1974-75 to 143 percent in 1981-82 due to changes in crop-plans as follows:
    i. Fallow - wheat was replaced by soybean + pigeonpea (inter-cropping),
    ii. Soybean – wheat (with limited irrigation), soybean – linseed, soybean – chickpea and soybean – safflower.
c. **Crop Yields**: A phenomenal increase in yields of *kharif* crops was observed as a result of improvement of external drainage of run-off water impounded against field bunds and increase in availability of well-water for *rabi* crops, introduction of new varieties and improved management practices (Table 14).

The data presented in Table 14 showed that local varieties of *kharif* crops also responded to improved drainage particularly in case of sorghum and soybean, but the response was limited, where as the improved varieties of various crops showed a remarkable increase in their yields with better management.

Table 14. Increase in yield of various crops (q ha^{-1})

Crop	Variety	Yield (q ha^{-1})		Percent increase
		1975-76	1981-82	
Sorghum	Local	4.0	5.9	47.5
	Hybrid (CSH-5)	6.8	18.7	175.0
Maize	Local	3.8	4.1	8.0
	Improved	8.6	22.1	157.0
Soybean	Black (Kalitur)	47.2	5.8	38.0
	Improved (JS2)	5.6	9.7	73.2
Gram	Local	5.0	7.5	50.0
	Improved	6.4	10.9	70.3
Wheat	Local	4.4	4.5	2.3
	Improved	5.6	8.5	51.8

d. **Remunerative crop- sequences:** With better rain–water management not only double cropping was a reality under rainfed conditions but also high yields of both *kharif* and *rabi* crops were obtained as shown in Table 15 (Verma, 1982).

Table 15. Remunerative Double Cropping Sequeces

S.N.	Crop sequences	Yield q ha^{-1}	
		Kharif	*Rabi*
1	Maize – Chickpea	35	10
2	Soybean – Chickpea	20	15
3	Sorghum (Green Fodder) - Chickpea	400	12-15
4	Sorghum early variety – Chickpea	35	12

e. **Effect of improved practices on yield:** Most of the farmers in Project area had no confidence in land management works due to unsatisfactory execution of soil and water conservation programmes earlier. Also, they did not readily agree to any works where water from neighbour's field could drain through the drainage channel passing through their holdings and the situation was the worst if two neighbours were not in good terms with each other. Further, some farmers over contended, typical of Kusum Nair's man with limited horizons, thought it too much of a botheration to adopt new technology. Hence it was necessary to demonstrate some easy practices to win their confidence, particularly the good effect of over - all - drainage on crop yields (Verma, 1980a) as presented in Table 16.

These results convinced the farmers that (i) removal of rain water stagnated against bunds was necessary for good yields of *kharif* crops, (ii) raising of *kharif* crops was not risky but more remunerative, if improved package of practices were adopted, (iii) yield of *kharif* crops could be increased by using fertilizers much more than that of *rabi* crops, and (iv) early sowing of *kharif* crops was more profitable.

Table 16. Effect of improved practices on yield

S.No.	Treatment	Average yield (q ha^{-1})		
		Sorghum CSH-5	Maize Chandan-3	Soybean T -49
1.	Traditional practices	6.8	5.5	8.8
2.	Removal of rain water standing against field/ property bunds, early sowing and application of low doses of fertilizers (20 kg N + 20 kg P_2O_5 ha^{-1})	23.0	13.5	10.5
3.	Removal of water stagnation, early sowing and application of recommended doses of fertilizer	32.0	27.5	12.2

f. **Crop Yields under phase 2 and 3:** The crop- yields recorded at the end of phase 2 and phase 3 (Pipliya Tapha watershed) were presented in Table 17 (Tomar, 2009)

Table 17. Yields of various crops (kg ha^{-1})

Crop	Local variety	Grain yield (kg ha^{-1})		Improved varieties	Grain yield (kg ha^{-1})	
		1987-88	1993-94		1987-88	1993-94
Sorghum	Vidisha 60-1	680	1430	CSH-5 SPV-475	1210	3780
Maize	Satha	390	1250	Ganga-5	410	2900
Soybean	Kalitur	320	850	JS-71-05, PK-472	480	1940
Wheat (rainfed)	Pissi local	280	-	Sujata	560	700
Wheat (irrigated)	-	-	-	WH-147	2200	4260
Chickpea	Ujjain-21	320	590	Annagiri	560	1840
Linseed	-	-	-	JLS-73-85	-	920

As was apparent from Table 17, the yields of even local varieties of various crops increased by 80 percent to more than 200 percent, as a result of better land and water management and addition of fertilizer, etc. The yields of hybrid/ improved varieties were increased by 25 percent to more than 600 percent. Minimum increase was in case of wheat where as maximum increase was in case of maize, as maize was very susceptible to water-logging.

g. **Fodder crops**: Area under fodder crops increased from 9 ha in 1974-75 to 80 ha in 1981-82.

h. **Milk yield**: Yield of milk (litres day^{-1}) per cow increased from 1.5 liters day^{-1} in 1974-75 to 3.0 liters day^{-1} in 1981-82 and per buffalo from 3.0 to 5.11 liters day^{-1} during the same period.

i. **Cooperative Society activated**: Over dues of J-H Cooperative Service Society were paid up and its supersession terminated.

4.9.4. Other impacts

a. **Improvement in water potential:** As a result of watershed development work, a significant improvement in underground water potential of the area was noticed (Table 18). In 1987-88, life span of shallow wells was only 2 to 3 months after the cessation of monsoon rains but in 1995-96 wells did not dry up till April. Also the irrigated area increased from 28 ha to 168 ha, by wells and construction of two tanks.

Table 18. Improvement in water potential (Tomar, 2009)

S.No.	Particulars	Base year 1987-88	Year 1995-96
1	Surface water storage (m^3)	Nil	27000
2	Total number of open wells	19	29
3	Shallow wells (water upto October-November)	3	27
4	Deep wells (water up to February)	6	6
5	Deep wells (live year round)	3	3
6	Tube-wells	3	30
	Live up to January	2	18
	Live up to April	-	8
	Live year round	1	4
7	Area irrigated (ha)	28	168
8	Irrigated area (percent)	6	40

b. **Better cropping systems:** Besides rise in water-table and increased crop-yields, there was increase in cropping intensity, increase in fodder production, and an increase in area under vegetables and floriculture. Also, soil loss from cultivated fields was reduced significantly and gullies were stabilized.

c. **Improved implements:** The number of improved blade harrows increased from 12 to 56, power threshers from 2 to 19, and bullock carts from 14 to 32. Previously there was no tractor but by 1995-96, there were 6 tractors belonging to farmers.

d. **Upgradation of livestock:** With increased fodder production, not only the number of buffaloes increased from 88 to 240 and number of cross-bred cows from 2 to 26 but also the milk production per buffalo increased from 4.8 to 6.2 litres day^{-1}, per cow from 2.5 to 3.3 litres day^{-1} and per crossed cow from 8.0 to 10.8 litres day^{-1}. Also, to start with there was no Gobar Gas Plant in the project area but by 1995-96 there were eight Gobar Gas Plants.

To conclude, watershed based development of rainfed farming brought lot of prosperity to all inhabitants and a significant improvement in over-all socio-economic condition of the farmers. This proved that watershed based development of rainfed farming was the appropriate approach and its technology paid high dividends.

4.10. OTHER PIONEER PROJECTS

After the grand success of Indo-UK Operational Research Project, the ICAR sanctioned 3 more pioneer ORPs on Watershed Development viz. Sukhomajri (Haryana), Fakot (Tehri Garhwal) and GR Halli (Karnataka). All these Projects made a mark in the development of rainfed farming through integrated watershed development as evident from their achievements.

4.10.1. Sukhomajri Project (Haryana)

The project was implemented in Shiwalik foot hills by Central Soil & Water Conservation Research Centre, Chandigarh (1975-79), mainly to check the rapid siltation (about 900 tonnes ha^{-1} $year^{-1}$ sediment) of Sukhna Lake through proper soil and water conservation measures. In first phase, a small earthen dam was constructed above a gully head. This checked the advance of gully and formed a farm pond with 0.8 ha m water storage capacity from a denuded hilly catchment of 4.3 ha. The water of farm pond was used to provide one supplemental irrigation to wheat crop and the yield increased from 16 to 42 q ha^{-1}. In the second phase, a watershed of 9.1 ha was treated with staggered contour trenches, check dams in gully and plantation of trees and grasses. Also, a 12m high earthen dam having water storage capacity of 5.6 ha m was constructed. This had miraculous effect in drought year of 1979. One life saving irrigation to maize produced a bumper crop and two irrigations of 7.5cm each to 16 ha wheat crop increased the average yield from 10 q ha^{-1} to 35 q ha^{-1}. Other important impacts were (CSWCRTI, 1980):

1. The denuded hill catchment of 16.1 ha changed into a green hill producing sustained yields of fodder and fuel.
2. Farmers produced quality fodder and other crops using improved seed and applying fertilizer.
3. Sediment from the catchment got reduced considerably.
4. Erosion in farmer's fields was reduced to negligible level.

4.10.2. Chitradurga Project on Integrated Watershed Management (1976-80)

An operational research project on integrated watershed management was launched by the ICAR on a red soil watershed of 314 ha in GR Halli village of Chitradurga district, Karnataka State during 1976-80. The watershed had 156 ha of denuded slopy hilly area and 158 ha of arable land. The treatment consisted of constructing a water diversion bund above arable land, check dam, drop structures and vegetative barriers at appropriate places. Staggered contour trenches on slopy hill area were constructed for in situ moisture conservation. Suitable plant species were planted on the whole hillock. The treatment of the watershed resulted in an increase of ground water recharge of about 23 percent. The irrigated area by wells in the watershed increased from 19 ha to 56 ha after treatment. The area under improved varieties increased progressively from 47 ha to 156 ha in 1994-95. Due to increase in irrigated area, improved varieties and other package of practices, the total production from irrigated area increased from 13 to 117 tonnes per year. The biomass production from non-arable land increased from almost nothing to 6 tonnes ha^{-1} $year^{-1}$. The economic well being of the people improved considerably after implementation of the watershed development programme.

Due to soil and water conservation measures, the annual run-off was reduced from 17 to 2.3 percent and soil loss to 1.1 tonnes ha^{-1} $year^{-1}$ (Sharda et al 2006).

4.10.3. Fakot watershed (Tehri Garhwal, Uttarakhand)

This watershed management project was located at village Fakot of Tehri Garhwal district in U.P. (now Uttarakhand) in lower and middle Himalayas. Project was implemented by Central Soil and Water Conservation Research and Training Institute, Dehradun, during 1975-88 in a watershed of 370 ha. The general slope of the area was 72 percent. Of 370 ha, about 82 ha was reserve forest and 80 ha agricultural land. Following works were undertaken:

1. Tanks: Five new cement lined tanks and 17 LDPE lined tanks were constructed and 2 cement lined tanks were renovated.
2. Guhls: 2310 guhls were cement lined, five new guhls were developed and one old guhl was renovated.
3. Terraces: 7.4 ha of irrigated terraces were renovated and 12.4ha of rainfed terraces (outwardly sloping) were converted into irrigated terraces (0.5 percent inward sloping).
4. Plantation on 36.6 ha waste-land: Horticultural plantation on 14.2 ha and fuel-cum-fodder plantation on 22.4 ha.
5. Plantation on terraces (rainfed) 7.6 ha.
6. Other works: (i) land-slide control works on 2 ha; (ii) Trail erosion control by stone check dams and water diversion drains 6 km long.

Result

i. Irrigated area increased from 11.8 ha in 1974-75 to 23 ha in 1985-86.
ii. Cropping intensity increased to more than 170 percent.
iii. Agriculture production increased 4.5 times.
iv. Area under horticulture plantation increased from 0.5 ha to 18 ha.
v. Fruit production increased from almost zero to 2408 q.
vi. Farmers earned Rs 7 lakh by selling flowers and vegetables.
vii. Fodder yields increased from 13t/ha to 36.8 t ha^{-1}
viii. Employment locally generated reduced migration from 26.7 to 9.3 percent.

The progress continued even after project was over as all farmers were actively involved in the implementation of the project.

4.11. GROWTH OF WATERSHED PROGRAMMES

Watershed management approach came into sharp focus after mid-seventies when eye opening achievements of integrated watershed development were obtained in the above discussed four pioneer operational research projects. Consequently, in 1983, I.C.A.R. in collaboration with State Governments and State Agricultural Universities selected 47 model watersheds for development in various agro-climatic regions. Further encouraged by good results of watershed based technology of rainfed farming development, the approach was adopted in various programmes funded by national and international agencies (World Bank, DANIDA, EEC (Now EU), Govt. of Germany, Swiss Govt. and JICA), e.g. Drought Prone Areas Programme (DPAP), Desert Development Programme (DDP), National Watersheds Development Programme for rainfed areas (NWDPRA), Integrated Watershed Development Project (IWDP), Western Ghats Development Programme and Hill Area Development Programme (HADP). During the last two years of Fifth Five Year Plan, emphasis was laid on integrated watershed development in 276 watersheds, as part of the soil conservation programme (Shankarnarayan and Shankar, 1984). Apart from these, various State Governments and NGOs also implemented watershed development projects in nineteen eighties and nineties. Some of the NGOs (Yeshwant Krishi Gram & Panlot Vikas Sanstha, Hiware Bazar, Distt. Ahmednagar (M.S.); National Centre for Human Settlements & Environment, Bhopal (M.P.); Centre for Sustainable Agriculture and Rural Development, Gwalior (M.P.); State Govts. of M.P., Chhattisgarh, U.P. and ICAR Institutes did commendable work on water-shed development which was recognised at national level. The National Productivity Council, Indian Farmers Fertilizer Cooperative (IFFCO) and Krishak Bharti Cooperative (Kribhco) gave national awards to Mettemari watershed (Karnataka). Chavela watershed (A.P.), Nartora watershed (Raipur, Chhattisgarh), Tejpura watershed (Jhansi, U.P.) and Ariel watershed (Bareilly, U.P.) Some of these very successful projects are discussed in continuation.

4.12. PROJECTS IMPLEMENTED BY STATE AGRICULTURE UNIVERSITIES

ICAR funded projects implemented by Jawaharlal Nehru Kristi Vishwa Vidyalaya (J.N.K.V.V.) Jabalpur, Madhya Pradesh: As extension of Indo-UK Project, following projects have been executed by Jawaharlal Nehru Krishi Vishwa Vidyalaya, Jabalpur, Madhya Pradesh.

4.12.1. Hingonia Pipliyatapha Project, Indore Campus (1987-88 to 1993-94)

As already discussed, the same type of land and water management works as in Indo-UK Project, were under taken viz. stabilization of main gully through loose boulder structure/gabions, construction of water diversion bunds, waste-weirs, spurs and two water harvesting tanks (10,000 and 17000 m^3 capacity). More underground water also became available due to rise of water-table as a result of watershed development work. Consequently, area under irrigation increased from 22 to 56 and from 28 to 168 ha in command areas of tanks. The cropping intensity increased from 133 to 165 percent. Soybean was introduced on large scale. The degraded land was put under silvi-pastoral system to increase fodder production. The milk production increased by 30 percent. The overall prosperity of the farmers and labourers increased. Also, due to watershed development programme, run-off and soil loss decreased from 30 to 17 per cent and from 1.5t/ha/year to 0.35 t ha^{-1} $year^{-1}$, respectively.

4.12.2. Berkheda Watershed Development Project, Distt. Guna, M.P.

It was executed by J.N.K.V.V. scientists to transfer the technology of land and water management evolved under Indo-UK Project, Indore, to Vindhyan Plateau Zone having medium black soils. Very good job was done by the Project.

4.12.3. Operational Research Project, Sidhi (M.P.)

Spread the same technology in Kymore Plateau Agro-climatic Zone, and wonderful results of run-off collection and recycling were achieved.

4.13. PROJECTS IMPLEMENTED BY CENTRAL SOIL & WATER CONSERVATION RESEARCH AND TRAINING INSTITUTE, RESEARCH CENTRE, DATIA (M.P.)

4.13.1. Bajni Watershed, Datia District, M.P.

The work was done in 1990-91 under Integrated Wasteland Development Programme (IWDP) funded by Ministry of Rural Development, Govt. of India, on 532 ha of land having both red and black soils. Red soil is gravelly, light textured and prone to erosion due to slope >5 percent. For channel stabilization, loose boulder and gabion cheek dams and bunds with waste-weirs on agriculture land were constructed. Silvi-pastoral, agri- horticultural and agro-forestry systems were adopted according to type of land. Desilting of village pond for renovation induced recharge of underground water and this when used through open dug wells or bore-wells improved the productivity of farm land. Crop diversification for fodder and fuel production increased milk yield and resulted in higher income generation.

4.13.2. Rendhar Watershed (Jalaun, U.P.)

An operational Research Project was executed on 748 ha watershed in Rendhar village during 1983-84 to 1991-92. The Project area was badly eroded into shallow and medium ravines by 5 *nalas* converging into the village. Out of 748 ha, 695 ha was arable and 639 ha was rainfed receiving an annual average rainfall of 803 mm. The pre-project productivity of various crops was very low, 4.5, 5.6

and 4.6 q ha^{-1} of sorghum, pigeonpea and pearl millet, respectively and 10.0, 6.2, 6.5, 2.5 and 5.8 q ha^{-1} of wheat, chickpea, barley, mustard and lentil, respectively.

Land smoothening on 458 ha, contour bunding on 541 ha, gully plugging on 198, ha strengthening of existing bundies on 398 ha, water harvesting bodies (47 ha), Cheek dams for water harvesting 25 ha, were constricted. Surfaces water storage of 1,80,000 m^3 was created.

To make better use of land, conservation forestry (93 ha), agro-forestry (17 ha), farm forestry (160 ha) and pasture (16 ha) were introduced. Improved package of practices for growing various crops resulted in 6 to 8 times increase in total production of the area. The productivity of crops during 1991-92 for sorghum, pigeonpea and pearlmillet was 28.6, 22.3 and 14.7 q ha^{-1} respectively and for wheat, gram, barley, mustard and sesame was 48.7, 27.3, 42.8, 26.7 and 19.5 q ha^{-1}, respectively. The production of milk and meat increased tremendously from 3,52,000 litres and 55 kg, respectively, during 1983-84 to 2292,000 litres and 10600 kg, respectively, in 1991-92. The employment generation was 1,10,000 man-days for landless labourers and 94800 man-days for farmers.

4.14. PROJECTS IMPLEMENTED BY STATE GOVERNMENTS

4.14.1. Khariya *nala* watershed (Mauranipur District, U.P.)

The total area of the watershed was 5395 ha, of which 3344 ha (62 percent) was cultivated land, 1675 ha (31 percent) community pastures (consisting of plateau, hill slopes sand rocky area) and 376 ha, (7 percent) wasteland. The soil and water conservation works started in 1990-91 included water diversion bunds, contour bunds, gully plugs and drainage outlet structures, 17check dams were constructed on *nalas*, which impounded 68 ha m water for irrigation. Also, 20 gully plugs impounded 10 ha m of water which helped recharge the ground water. The dug out wells increased from 82 to 930. The irrigation potential increased from 9.6 to 83 percent of the cultivated area and consequently cropping pattern and cropping intensity changed from 81 per cent to 192 percent. Sorghum was replaced by groundnut, soybean was introduced and area under tomato increased from 6ha to 595 ha. Wheat area increased from 490 to 1720 ha (3.6 times). The productivity of *kharif* crops increased to 1.7 t ha^{-1} from 0.6 t ha^{-1} and *rabi* crops from 0.8 to 3.5 t ha^{-1}. Availability of fire wood and fodder and carrying capacity of pasture land increased enormously.

Run-off and soil loss decreased continuously soil loss decreased from 41 t ha^{-1} to 18.5 to 1.0 ha^{-1} finally in 7th year. The run-off from hillocks, waste-land and arable land got reduced from 70 to 19 percent, 48 to 15 percent and 30 to 18 percent, respectively. The water table rose by 4 to 6 m over years.

The landless families got increased income from basket making by cutting raimunia (*Lantana camera*) bushes and by preparing plates from dhak (*Butea frondosais*) tree leaves.

4.14.2. Nartora watershed (Raipur district, Chhattisgarh)

Integrated Watershed Development Plan was implemented on 704 ha land in Mahasamund block of Raipur District (Chhattisgarh State), having lateritic soil and 1 to 3 percent slope. Of 455 families in the watershed, 83 percent were small and marginal farmers. Sixty nine percent of total area was arable, nine percent forest 15 percent fallow and 6 percent under pastures. The land treatment consisted of bunding and smoothening where needed. Three water-harvesting tanks were constructed/rejuvenated which provided irrigation to 95 ha. Additional net cropped area of 16 percent and increase in cropping intensity of 15 percent was obtained. Yields of paddy, groundnut and wheat crop were increased by 25, 32 and 12 percent, respectively. The overall net income per farm was increased by 88 percent.

4.14.3. Parua *nala* watershed development project (Sehore district, M.P.)

The Project was implemented by Department of Agriculture, M.P. Government during 1982-87, with World Bank funds. The whole area was divided into 4 milli- watersheds of about 10,000 ha each on both sides of main Parua *nala* and these milli-watersheds were the catchments of 4 sub *nalas* joining the main *nala*.

Management of land resources under the watershed programme included cultivated as well as uncultivated land under the ownership of farmers, Panchayats and the government: The land treatment consisted mainly of levelling and bunding for diversifying the existing farming systems resulting in improving the productivity of not only agriculture and allied commodities but also the overall production of biomass for enhancement of self employment opportunities and thus the overall income of the rural household. An attempt was also made to restore the ecological balance in the degraded and fragile rainfed ecosystems by greening these areas through appropriate mix of trees, shrubs and grasses. This resulted in creation of sustained employment opportunity for rural community including the landless and ultimately the regional disparity between the rainfed and irrigated areas was reduced (Kushwaha, 2012).

4.14.4. Integrated watershed management project for ravineous areas of Chambal and Yamuna catchments (Agra and Etawah districts of U.P.)

The project became operational in March, 1986 and lasted till March 1993. It was very big project funded by European Economic Commission (EEC) now European Union (EU). The project area was 47,420 ha extending upto confluence of Chambal and Yamuna, divided into nine macro-watersheds. The project was executed by five departments of Government of U.P. *viz.* (i) Soil Conservation Service, Department of Agriculture, (ii) Forest department (iii) Nal Koop Nigam, (iv) Rural Engineering Service, and (v) U.P. Electricity Board.

4.14.4.1. ***Land Development Works:*** The soil conservation service executed various land development works presented in Table 19.

Table 19. Land Development Works

S.No.	Type of lands	Originally targeted area (ha)	Actually covered area (ha)	Activities
1.	Table land (slope <1percent)	5400	5877	Field bunding with pipe outlets for disposal of run-off from each field
2.	Peripheral land	7850	12503	Construction of peripheral bunds; Smoothening, leveling with water disposal structures.
3.	Shallow ravines (depth <2.5m)	13750	13739	Terracing with run-off disposal structures (pipe culverts, chutes, etc.
4.	Medium and deep ravines (depth 2.5 to 5.0m)	20430	16522	Stabilization of ravines through construction of gully plugs, check dams, waste-weirs, culverts, etc. and afforestation.

4.14.4.2. ***Utility plantation:*** On community land (4259 ha) utility plantation to get fuel wood, fodder timber and fruits was done. This also provided a vegetative cover against soil erosion. The overall density of plant population was 1044 plants ha^{-1} and the average survival rate was 48 percent. Besides, afforestation on 10,000 ha of deep ravines was done planting deshi babool (*Acacia nilotica*), bilayati babool (*Prosopis juliflora*), shisham (*Dalbergia sissoo*), cypress (*Capressus sempervirens*), ber (*Ziziphus jujube*) etc.

4.14.4.3. ***Crop production programme:*** Demonstrations were conducted on already popular crops as well as on more remunerative crops such as sunflower and soybean to popularize new crops and improved varieties and package of practices for various crops. The programme was conducted on 35450 ha benefitting 32810 farmers.

4.14.4.4. ***Tube-well Construction:*** To tap 67 MCM of available under ground water, 195 tube-wells were constructed. The tube-well system consisted of a pump house, an elevated tank, a pumping unit with electrical equipment, and an underground water distribution system (for 50 ha) having 160 mm diameter PVC pipe-line, with outlets at appropriate sites. The total cost per tube-well system was Rs. 12.54 lakhs. The tube-wells were constructed by Nal Koop Nigam and energized by U.P. Electricity Board.

The irrigation potential created was 14100 ha but underground water distribution was on 7050 ha only and the potential really utilized was 4476 ha (31.8 percent).

4.14.4.5. ***Link roads construction:*** The project area was most backward from communication point of view. Since roads play an important role in the development of an area, provision was made in the project for the construction of link roads. It was possible to cover larger lengths of non-metalled chak roads with the provision made in the budget but Rural Engineering Services Department, Soil Conservation Department and representatives of villagers and village leaders all wanted to connect project villages to the main roads/highways through metalled roads. Accordingly, 60.3 kms. roads and 149 culverts were planned by Rural Engineering Services Department - 40.45 kms. in Agra District and 19.85 kms. in Etawah district during 1989-90 to 1992-93. The total cost of roads and culverts was Rs.41.71million averaging to Rs. 6.92 lakh per km. With construction of roads, an era of happiness and prosperity dawned in the project area and all the villagers had a sense of social up gradation.

4.14.4.6. ***Project benefits:*** Following benefits of the project were apparent:

1. Employment generation: 11.84 lakhs man-days employment was generated.
2. Additional area for cultivation: 13040 ha of additional land created for cultivation.
3. Additional irrigation potential created: 19,500 ha
4. Additional Income:
   a. Increased agricultural production through increased area and productivity was Rs. 917.13 lakhs $year^{-1}$.
   b. Marketing of produce at more competitive rates (due to roads).
   c. Increased supply of fuel wood, fodder and timber
   d. Increased employment of farm labour due to increase in cropping intensity and more intensive agriculture.
5. Economic Rate of Return from Tube-well irrigation: 25 to 30%.

4.14.4.7. ***Evaluation:*** The technology adopted was very expensive and it appeared as if it was a construction project. Pipe outlets caused erosion in fields and therefore chute spillways would have been better. However, farmers and other inhabitants were happy due to construction of roads for obvious reasons (WAPCOS, 1993).

4.15. PROJECTS IMPLEMENTED BY NGOS

4.15.1. Hiware Bazar Watershed Project (District Ahmednagar, M.S.)

The Project was executed by Yeshwant Krishi Gram & Panlot Vikas Sanstha, Hiware Bazar, during 1993-94 to 1999-2000 (YKGPVS 1991). The project area was 977 ha of which 795 ha was cultivated land, 170 ha wasteland and 7 ha pasture land. Average annual rainfall was 350 mm and the area had shallow, medium and deep black soil. The irrigated area was 360 ha by 177 wells. The forest and wasteland had bare hillocks and slopy land (3-25 percent slope).

4.15.1.1. Development works: In 1993-94, six lakhs saplings of deshi babool (*Acacia nilotica*), gliricidia (*Gliricidia sepium*), sisham (*Dalbergia sissoo*), etc. were planted in contour tranches on bare hillocks and wasteland. Tree plantation on private land 124 ha, on field bunds 206 ha and on road side 6 km was also done. Contour bunding was done on 414 ha, *nala* bunding at 51 spots on one main *nala* and two secondary *nalas*, 4 submersible check dams and 7 masonry dams with wooden planks on main *nala*, two percolation tanks to catch run-off from hillock, 12 loose boulder structures and two earthen plugs pitched with boulders were constructed as soil and water conservation measures.

4.15.1.2. Crop Plans: The area mainly produced sorghum, pearlmillet and pulses and had 5 ha land under orchards, 4 ha under vegetables and fodder crops and 1 ha under oilseed crops before project implementation.

4.15.1.3. Results: Although total annual rainfall was 350 mm, but heavy storms for short duration were common e.g. in one year, 120 mm (1/3rd of total) rainfall was received in 24 hours. This resulted in severe erosion, as was evident from big silt deposits against all structures. The silt deposit led to stabilization of *nalas* and run-off collection resulted in rise of underground water table. The number of live wells increased to 190 from 120. This led to increase in summer irrigated land from 10 ha to 100 ha. Farmers started growing onion, potato, cotton, vegetables and flowers instead of sorghum and pearlmillet. The area under vegetables increased to 25 ha, pulses to 45 ha, oilseeds to 5 ha, fruit orchards to 35 ha and fodder crops to 50 ha. Total agricultural production of the village rose to Rs. 45 lakhs year^{-1} from Rs. 9.45 lakhs year^{-1}, while per capita income to Rs. 4000 from Rs. 832 per year^{-1}.

4.15.1.4. Watershed Development to Rural Development: The active interest of Gram Sabha changed this Watershed Development Programme into Rural Development and social uplift. It put a ban on grazing and tree felling. This changed the dark bare hillock into lush green hill slopes, yielding increasing quantities of fodder every year. The fodder production from hill slopes increased to 1600 tons in 1999-2000 from 200 tons in 1994-95. The problem of shortage of fuel-wood due to ban on tree-felling was solved by biogas plants. About 50 biogas plants operated successfully in the project area.

The Gram Sabha banned flood irrigation to banana and sugarcane and introduced drip and sprinkler irrigation. Gram sabha introduced prohibition in village having 6 illicit liquor dens in 1990. It encouraged people for increasing milk production which rose to 2000 litres day^{-1} in 1999-2000 from 300 litres day^{-1} in 1994-95. There was only a primary school in village offering education from 1st to 4th standard. It changed into Zila Parishad Middle School (up to 7th standard) three years later (1997-98) through donation of land, money, etc. by the villagers. Women Self Help Groups were organized for poultry, sheep keeping, etc. Poultry farming and sheep/goat farming generated employment and income to members of self help groups. The Gram Sabha has taken care of environment protection, health and sanitation. Health examination of villagers was arranged twice a year. The villagers are all healthy and happy and no epidemic-break out took place.

4.15.1.5. Awards: Several awards were won by the project for its excellent work, e.g.

1. Lokmat Puruskar 1997 for environmental and social work
2. Adarsh Gav Puruskar (Ideal village prize) for best rural development
3. National Productivity Award 1996-97 by Govt. of India for best work on Dry Framing.

4. Vanshri Puruskar 1998-99 by Government of Maharashtra for afforestation.

The Times of India dated April, 21, 2001, carried a story of the project entitled "In Drought hit Ahmednagar-an oasis".

4.15.2. Jhabua watershed development Project (Distt. Jhabua M.P.)

The Project was executed by National Centre for Human Settlements and Environment (NCHSE), Bhopal, M.P. during 1991-94 (NCHSE 1995). The population involved was 3084 in 558 households of 6 villages of Jhabua District and comprised of mostly tribals (85 percent). The area of watershed was 2421 ha of which 507 ha was degraded (highly eroded) land. Such area was increasing at a fast rate due to deforestation, overgrazing and inappropriate land-use. Increase in population forced people to convert forest and grazing land into agricultural land, but the land was not able to sustain human and livestock population. Consequently, people were migrating to other places in search of employment.

4.15.2.1. Works done: To adopt integrated watershed development approach, soil and water conservation measures both engineering and biological and appropriate land use according to land capability classes was practiced. The local community (tribals) was fully involved in all programmes at every stage i.e. formulation, implementation and management of programmes. To start with 310 ha of degraded land was treated for soil and water conservation by contour-trench, gully plugging and construction of gabion structures. During 1992, silvi-pastoral system was developed on 80 ha of community and private land and it was extended to 507 ha of land in 1993-94. Seven farmers' nurseries were established to grow 5 lakhs saplings of bamboo (*Bambusa glaucophylla*), desi babool (*Acacia nilotica*), siris (*Albizia lebbek*), mango (*Mangifera indica*), guava (*Psidium guajava*), etc. One masonary stop-dam of 21.5 m length and 2.0 m height with gross storage capacity of 16,438 m^3 was constructed.

4.15.2.2. Results: As a result of soil conservation works, land productivity increased. Gross production of 50 metric tons of fodder was obtained in the first year from 80 ha of degraded land where not a single blade of grass was growing previously. The village committees decided how the product was distributed among the participants. The stored rain water was used for domestic purposes and it helped in recharging the ground water. The people got enough forage grass and fuel wood.

About 600 smokeless and portable fuel-efficient stoves were distributed. As a result, large quantities of fuel wood were saved and the environment was also kept clean. A much greater awareness was observed among the people to reclaim degraded land, manage common property and protect the environments.

4.15.2.3. Award: The project won 'Desertification Control Award' of Government of India.

4.15.3. Watershed based reclamation and control of ravines, Noorabad, (District Morena, M.P.)

The project was executed by Centre of Sustainable Agriculture and Rural Development, Morar, Gwalior, on farmers' fields at Noorabad village (District Morena) on National Highway No.3 about 30 kms from Gwalior (CSARD 1998). Since it was first attempt to reclaim and control ravines through integrated watershed management, a microwatershed of about 100 ha was selected for development to start with. Of this, 35 ha was table-land, 50 ha under shallow and medium ravines and 15 ha under deep ravines.

4.15.3.1. Work done earlier: During 1970-84, ravine reclamation work was undertaken under central sector scheme by soil conservation service, Department of Agriculture, M.P. Government. Noorabad and Tigara ka pura villages were included in the programme. Field bunding in table-land, levelling of shallow and medium ravines and masonary structures of different kinds (walls, submersible check dam, etc.) were constructed. No treatment of U-shaped and broad based deep ravines was done, as a very strong and long submersible masonary check dam was constructed in early nineteen forties by Department of Agriculture, Gwalior State Government. While this stabilized the bed of deep gully it remained submerged under water during the rainy season and crop was grown in *rabi* season (mustard mainly) on the bed only. No attempt was made to utilize banks which had stabilized taking a natural angle of repose. Also, the levelled land (previously under shallow and medium ravines) started developing rills, washes etc. due to run-off as masonry structures were inappropriate (simple walls, etc.) and mostly at inappropriate sites (between two earthen bunds which were not stable). While walls and other structures stood intact, run-off passed beyond the structure through side/sides defeating the purpose of structure construction. Thus, soil conservation works resulted only in temporary reclamation of ravines and no control on their formation.

4.15.3.2. Climate: The climate of the project area was semi-arid with mean maximum temperature of 31.3°C and mean minimum temperature of 18.2°C in summer. The average annual rainfall was 675 mm of which 624 mm was received in July-September. However, much higher annual rainfall was received in 1995 (1213 mm) and in 1996 (1105 mm).

4.15.3.3. Soil: Typical alluvial soil (Ustochrept) yellowish brown to dark yellowish brown; deep, fine sandy loam; neutral to slightly alkaline; with very high dispersion coefficient.

4.15.3.4. Agro-economic situation: There were 98 farm families and small and marginal farmers dominated. Upper two third part of the watershed was rainfed while lower one third had well irrigation. Main crops grown were pearlmillet and pigeonpea in *kharif* and mustard and wheat in *rabi*. Only one crop a year was grown following *kharif* – fallow – fallow - *rabi* rotation. Crop yields were low due to local varieties (usually long duration ones) and no use of fertilizer. Annual income per house-hold was Rs. 32000/- (in 1995) of which Rs. 9,200 (on average) per annum was from employment in factories of Banmore Industrial area and Rs. 11000 per annum from milch animals.

4.15.3.5. Works executed: Works were executed for stabilization of main gully, side gullies and smoothening of land (where necessary).

a. **Construction of tank:** A tank was constructed at the junction of first two sub-gullies joining the main gully by construction of a bund across the slope (3 m at the base and 2m high) having 1:1 side slope upstream and 2:1 on the other side.

b. **Gully plugs:** Six earthen gully plugs on the main gully and seven gully plugs on side gullies were constructed. All gully plugs on the main gully and two plugs on the side gullies were provided with gabion-waste weirs as surplussing arrangements for the run-off.

c. **Land smoothening**: Smoothening of earthen mounds (hummocks) and narrow long and high banks of land, formed by side gullies/ravines, was done, so as to broaden the main gully and make more area available for cultivation.

d. **Construction of gabions:** In all 9(nine) gabions were constructed, 3 (2+1) on side gullies - two on a long side gully and one on other side gully. Six gabions including one as waste-weir of the tank were constructed on the main gully - one at each gully plug on the lowest point. G.I. wire (14 gauge) was used to knit nets for upper 2 gabions and 12 gauge G.I. wire for remaining gabions. The gabions were made with simple shoulders (on both sides) so as to make a rectangular weir above the crest-wall. An apparon (mattress) was provided along the whole length of crest-wall (including shoulders). In all gabions, 40 cms deep foundation was provided. A reverse filter consisting of coarse sand and gravels (<1.5 cms) layers – 10 cms thick each, was provided. Since it was the first time when gabions were used for gully-bed stabilization on ravineous alluvial soil, different heights of crest-wall were tried for the satisfactory performance of gabions. The theoretical height of a crest-wall of the gabion was equal to difference between bed levels of mattress of one gabion and height of the foundation-wall of the next gabion i.e. bed level at the point where next gabion was being constructed. Also reverse filters below shoulders and crest-well and against the crest –wall upstream were tried to test these.

4.15.3.6. Crop-Demonstrations: Crop demonstrations on improved and high yielding varieties of mustard, pearlmillet and pigeonpea were conducted. Pusa bold variety of mustard, yielded 8.5 to 11.0 q ha^{-1} as compared to previous average yield of 5 to 6 q ha^{-1}. The increase in yield was due to balanced use of fertilizer and deposit of silt on soil, and check of soil erosion. Pearlmillet variety Vijai composite yielded 18 to 24 q ha^{-1}, while pigeonpea variety JKM–7 produced 12 to 14 q ha^{-1}. Also, previously farmers were growing pigeonpea and pearlmillet seeds mixed together and broadcasted. They were persuaded to follow intercropping system (2 rows of pigeonpea and 4 rows of pearlmillet) which yielded about 16 q ha^{-1} pearlmillet and 12 q ha^{-1} of pigeonpea. This not only made farmers to adopt intercropping but also encouraged line-sowing instead of mixed sowing.

4.15.3.7. Results of gabion technology: Based on the successful performance of gabions to stabilize and control gullies on black soils tried under Indo-UK Project, nine gabions were constructed in the present project area including one as tank-waste weir. While gabions on side gullies worked very satisfactorily, some gabions on earthen gully plugs constructed across wide U-shaped gully had problems of breaches on one or either sides of the gabion, because the soil of gully plugs was not compact, nor the gully plug was stabilized by vegetative cover. Pitching of sides of the gully plug (earthen) also did not work. Those gabions whose both shoulders were fixed into consolidated gully-walls worked satisfactorily. In such a soil, the gabion made of dressed boulders worked better than irregular shaped ones as the latter ones could not be packed well to avoid making the structure too much porous. As the structure had big pores althrough, the velocity of run-off was not reduced enough to completely unload sand/silt particles carried by run-off. Also, gabions having deeper foundation (60 cms) worked better when crest-wall height was one metre or higher. Also, the foundation should be deeper-at least 45 cms below the original flow line of the gully. As regards reverse filter, it should be at least 10 cms thick layer of sand followed by 10cms layer of black gravels *gitti* (size about 15 mm) for every one metre of crest-wall. The reverse filter must be placed in foundation of gabion including shoulders as well as below mattress.

4.16. CONCLUSION

To conclude, watershed based development of rainfed farming was the only appropriate approach to exploit its huge potential. This approach brought lot of prosperity to all inhabitants and a significant improvement in over-all socio-economic condition of the farmers. The technique of integrated watershed management was very successful not only on cultivated land but also on ravineous waste-lands as demonstrated on farmers' fields in European Economic Commission (now European Union) and Department of Science and Technology, Govt. of India, funded projects in Agra-Etawah districts (U.P.) and Noorabad, District Morena, M.P., respectively.

■■■■

V

LAND AND WATER MANAGEMENT PRACTICES

Under rainfed/dryfarming conditions, availability of adequate moisture throughout the plant growth period was the limiting factor for crop-production, even in high annual rainfall regions, due to erratic distribution of rainfall, high potential evapo-transpiration and the limited soil moisture storage governed by texture and depth of the soil. Therefore, main strategy of rainfed farming development was the appropriate management of rain-water on watershed basis as it was the only source of water for rainfed crops. Accordingly, the technology for watershed development consisted of mainly (a) Land and rain-water management practices, and (b) crop husbandry practices. The appropriate land and water management practices aimed at the most efficient use of land and water resources. The land was to be put to most productive use for which the land was capable without causing its degradation whatsoever. Similarly, most judicious use of available rain water was also to be made. The rain water when started flowing as surface runoff and the flow velocity exceeded the permissible non-erosive velocity caused severe soil erosion. If it moved at too slow a rate on flat lands, it resulted into water stagnation/water-logging. In fact the runoff water must always flow with a certain range of velocity, depending upon the soil and surface characteristics which was practically almost impossible at all places. Hence the flow velocity of rain water led to either or both the problems in a catchment and to overcome these, appropriate land and water management practices were to be adopted. The practices should lead to safe disposal of excessive runoff without damaging the land. This means the land has to be managed to manage rain-water depending upon the nature of the problem created by it, i.e. soil erosion due to excess slope (greater than one percent) or water-logging on land having less or almost no slope (less than one percent) . Therefore, land and water management go together.

5.1. IN-SITU WATER CONSERVATION

Soil and water conservation mainly consisted of reducing the run-off and thereby check/minimize the soil loss. Creation of vegetative cover during rainy season was the most effective and the cheapest measure for in situ water-conservation. Planting of grasses on slopy land (up to 3 percent slope) and contour farming are the easiest and most effective low cost methods to reduce the rate of run-off and allow maximum infiltration of rain-water into the soil. Planting of crops along the contour lines on permeable soils (Entisols, Alfisols, Inceptisols) and on 0.2 to 0.3 percent grade on impermeable/less permeable soils, created a suitable barrier against free flow of run-off, dissipated the energy of flowing water and provided greater opportunity for run-off absorption/ infiltration into the soil. This not only reduced the amount of run-off but also reduced considerably the erosive velocity of run-off leading to increased crop-yield, less soil-loss and greater recharge of underground water (Verma et al 1981).

5.1.1. Efficiency of different cover-crops

The efficiency of a cover-crop depended upon its rate of canopy development, particularly because initial rain period of 30 days was very crucial from erosion point of view due to high intensity of rainfall and less crop-canopy during this period. Studies conducted at Central Soil and Water Conservation Research and Training Institute, Research Centre, Datia, revealed that run-off from cultivated fallow land with two percent slope was 45 percent of the total amount of initial rainfall received and a good crop cover could bring it down to about 20 percent. This proved the effectiveness of crop-cover in reducing run-off during initial period of rains (Tiwari and Narayan 2013). Further, cultivated legumes such as cowpea, blackgram, greengram, clusterbean, groundnut and soybean developed canopy at a fast rate and were consequently considered as better cover crops than sorghum, pearlmillet and maize. In an experiment conducted again at Datia, three legumes viz. greengram, blackgram and cowpea, were identified as promising cover crops for the region, yielding 6.0 to 8.0 q of grain ha^{-1}. In another study conducted on a slopy red soil, groundnut also emerged as an efficient cover crop which recorded the minimum soil loss to the tune of 3.6 tonnes ha^{-1} as against 8.6 tonnes ha^{-1} under cultivated fallow plot (Tiwari and Narayan, 2013).

5.1.2. Other practices

On semi-arid and sub-humid Vertisols, soybean has been found to be a very suitable cover crop. Its effectiveness as cover crop on land with about two percent slope increased considerably when planted across slope. On land with three percent slope, soybean planted across slope with barrier of lemon grass (*Cympopogan martini*) was very effective in reducing run-off whereas on land-slopes 3 to 6 percent, soybean planting across slope along with contour/graded bunds (earthen) resulted in reduced run-off and soil less and consequent high yield of soybean, as shown in Table 20 (Jain 2012).

Table 20. Effect of vegetative cover, hedge barriers and bunds on run-off, soil loss and yield of soybean

S.No.	Treatment	Mean over years 2002-08		
		Yield (kg ha^{-1})	Run-off (mm)	Soil loss (Tonnes ha^{-1})
1	Planting along slope	1539	80.6	2.31
2	Planting across slope	1725	44.7	1.01
3	Planting across slope + vegetative barrier	1841	32.8	0.73
4	Planting across slope + contour/graded bund	2211	25.4	0.60

All these measures increased infiltration of water into soil and consequently the in-situ moisture conservation.

5.1.3. Use of Mulches

Various types of mulches have been found to be very effective for in-situ moisture conservation. Various materials spread on soil surface as mulch dissipate the kinetic energy of rain-drop and thus prevent splash-erosion. Further, these materials obstruct run-off and consequently both rain-water and soil particles stay in place. Not only this, these conserve soil moisture, control weeds, increase infiltration of rain-water into soil, reduce run-off, increase charging of underground water and ultimately the productivity of crops where mulches are used. In a study conducted at Indore, polythene (ordinary grade), straw, weed biomass and glyricidia (*Gliricidia sepium*) green leaves were found very effective as mulches for in-situ moisture conservation (Jain, 2012).

5.2. LAND AND WATER MANAGEMENT TO CONTROL SOIL EROSION

Generally climate, topography, vegetation and soil type are some of the major factors affecting soil erosion. The soil erosion was the function of two properties namely erosivity of rainfall and the erodibility of land. There was no control over erosivity of rainfall - its time of occurrence, amount and the intensity. Therefore, the erodibility of soil/land had to be minimized by its appropriate management. The land use and crop management decided the degree of plant cover over it. Under natural conditions, the land was covered by vegetation. The vegetation cover not only protected the soil against erosion but it also increased the rate of infiltration of water into the soil which was particularly important on black soils as these become almost impervious when saturated with water. Also, in black soil region enormous storage of water on land surface developed under natural conditions due to gilgai formation, and consequently the run-off was reduced. But as soon as the land was brought under the plough, problems of soil erosion by rain water got started. On cultivated land, the soil erosion could be checked by either creating effective vegetation cover at a low cost or by altering the steep topography through very costly land shaping operations or by a judicious combination of the two. The topography and the soil and climatic conditions were considered imperative in deciding the suitable approach which should, therefore, aim at reducing the contribution of rain water to soil erosion by suitably managing the land, the soil and the plant–cover and at obtaining the maximum water use efficiency of rainfall by runoff collection and recycling, that is, the land had to be managed in such a manner that excessive runoff left the field without damaging the soil/land (Verma 1983). Obviously, various types of measures viz. biological, cultural and mechanical were to be integrated together to interact and provide a practical and economic way of conserving the soil and water resources, depending upon the land and soil characteristics governing the erodibility of the soil and the erosivity of rainfall (Verma 1988).

5.3. MANAGEMENT OF SLOPY LAND

Since the land slope played the most important role in determining the erosion hazards, the land management practices to be adopted for soil water conservation also depended upon the land-slope. The appropriate land management practices for different types of slopy land were as follows:

5.3.1. Bare and steep hill slope

On the ridge and upper reaches of a watershed, hillocks with slopy land around are quite common, particularly in black soil region. Under natural conditions hill slopes and associated lands were fully covered by trees and grasses. But consequent to reckless over exploitation of forests by human beings, occurrence of bare hillocks in the watershed was quite common. The slopy hillocks and adjoining land with slope greater than 15 percent was subjected to severe water erosion. The run-off water ran in a number of very small streams on the slopy land, loaded heavily with silt and finer particles (detached and dispersed by splash erosion caused by rain-drop) and picked up the speed fast on slopy land. Greater the slope, greater was the velocity of run-off water. At a certain point where the slope was less, the small streams converged into a bigger one, the rills turned into a shallow gully deepening with time. Due to severe erosion, there was hardly any soil left on the slope, and thus bench terracing on such land was out of question. The only way to stabilize bare hill slopes was afforestation of these. Planting of trees for fuel, fodder, fibre and fruits in contour trenches dug against the slope on hill slope had to be done. Planting in trenches was better than in pits dug for the purpose. On inter-trench area natural vegetation including grasses regenerated as some sort of protection was given to trees, at least in beginning. This was quite easy and practicable particularly if the inhabitants of the area were taken into confidence. As regards various fast growing tree species to be planted, the same (i) for fuel and fodder were: subabool (*Leucaena leucocephala*), shevri (*Sesbania aegyptica*), agasthi (*S. grandiflora*), bastor (*Albizia amara*), kalasirus (*A. lebbek*), safed sirus (*A. procera*), babool (*Acacia nilotica*), and nilgiri (*Eucalyptus hybrida*); (ii) for fruits were: Custard apple (*Annona squamosa*), guava (*Psidium guajava*), grafted plum (*Prunus cerasifera*), aonla (*Emblica officinalis*) and karonda (*Carrissa carandas*).

5.3.2. Steep land (slope >6 percent to 15 percent) with shallow soil

Such land usually occurred in foot – hill region and adjoining areas and was subject to severe soil erosion, particularly if natural grass cover was disturbed by overgrazing or by human effort to raise crops. If the slope of land was more than 6 percent and the soil was too shallow (less than 30cm deep) to undertake terracing/bench terracing, the land should not be used for arable farming but should preferably be diverted to alternate land-use such as silvi-pastoral or horti-pastoral system. The conversion of cultivated land to grass land will reduce erosion to at least one tenth, if not one hundredth of current rates (Hurni, 1988). The trees for fuel, fodder and fruits could be planted in trenches as already detailed. Various improved grasses which could be grown on slopy land to arrest soil-erosion were Dinanath grass (*Pennisetum pedicellatum*), hybrid Napier (*Pennisetum purpureum*), Guinea grass (*Penicum mascimum*), Blue pannic (*Penicum antiotale*), and Para grass (*Brachiasia mutica*). Besides providing an effective vegetative cover, grasses increase the infiltration rate and consequently reduce runoff. The practice of growing legumes such as stilo (*Stylosanthes hamata*), siratro (*Phaseolus atropurpureus*), pillipasara (*P. trilobus*), chipkani (*Desmodium contortum*), and sheora (*Alysicarpus rugosus*) with grasses resulted in not only more nutritious fodder but also minimized the use of nitrogenous fertilizer to pastures. A grassland after establishment had to be properly managed if good fodder, both in quantity and quality, was to be harvested from it, and if it was to protect soil against erosion. Continuous over grazing of existing grasslands resulted in denudation of the cover, increasing the rate of runoff and soil loss. The quality of pasture was preserved longer if the grass was cut and fed to animals rather than free grazing, particularly on black soils which got deteriorated much faster due to trampling by grazing animals. Under farmers' conditions, again, this could be ensured by active cooperation of all the inhabitants of the area.

5.3.3. Steep land (slope>6 percent)with medium and deep soil

Various situations under which such land occurred were:

a. Topo-sequence of shallow- medium-deep soils beyond foot-hill region in black, red or alluvial soil region, i.e. middle and lower parts of a typical watershed with red, black or alluvial soil.

b. River banks and adjoining areas the extent of which was much more in alluvial soil region than in black soil - ravineous land with medium and deep ravines.

c. Ill-managed and eroded land with undulating topography or land under shallow, and medium ravines following table land.

Management of steep land with medium and deep soil usually meant dealing with rates of run-off, rates of infiltration, rates of water percolation and extent of water erosion. There was lot of evidence that steep land could be dealt with successfully through appropriate management practices incorporated into a farming system, e.g. mulching with crop residues, adoption of conservation tillage, strip tillage and till- plant to increase infiltration and reduce run-off. If the land had a slope greater than 6 percent, the serious problems of soil erosion could not be controlled by vegetative cover alone. Also, under most cases effective vegetative cover could not be established under farmers' conditions before heavy rains hit the ground. Thus, adoption of appropriate mechanical measures was necessary on such land.

5.3.4. Mechanical measures

When erosion could not be controlled adequately by establishment of vegetative cover and other agronomic measures, adoption of mechanical measures was necessary. These measures included land smoothening, terracing, contour bunding, graded bunding, gully control measures, etc. and could be adopted under different situations. The contour/graded bunding controlled erosion by (i) intercepting the runoff and thus increasing the time of concentration on one hand and increasing the infiltration on the other; (ii) converting the long slope into several short ones and thus reducing the velocity of runoff. Though, land levelling and land shaping was often recommended as a soil conservation measure, it was too expensive to be afforded particularly under rainfed conditions. Further, even after levelling adoption of suitable measures to hold the soil in situ against water erosion was necessary. Before deciding the appropriate mechanical measures, it was necessary to have basic information like soil type, topography, location of gullies, *nalas*, and road, etc., frequency of floods, external drainage condition, drainability of the land, general crop yields and the value of agricultural land.

5.3.5. Terracing

The situation under which terracing is necessary and feasible existed over limited areas in Central India. Terracing the banks of deep ravines might not be viable because of very fragile nature of soil there, and high cost of maintenance needed for such terraces. Further, very steep slope of banks (>100 percent slope) of deep (depth about 20m) ravines and peculiar shape of soil columns (almost isosceles triangle with narrow to very narrow base, very long sides and acute top angle) between two very deep gullies, make it almost impossible to convert these soil columns into terraces. Terracing in not so deep but wide gullies situated away from one another could be a sound proposal. However, terracing was highly skilled job and needed constant expensive maintenance. The main constraint in the construction of terraces was the high labour input required to attain reduced slope gradients. Recent research has focused on developments whereby natural erosion, land management and/or animal power was used to build terraces over long periods of time, rather than direct construction. It had been shown that many of the techniques applied in direct construction were not suitable to local ecological and socio-cultural situations (Shaxson 1980). Level structures in high rainfall areas could not retain maximum runoff during heavy storms, especially on cultivated land. Livestock must be excluded from areas where terrace-farming practices were applied whether physical or biological. Also, there was a limit to terrace formation because of insufficient depth of soil and excessive slope gradient. These factors had to be considered at the design stage of conservation activities (Hurni 1988).

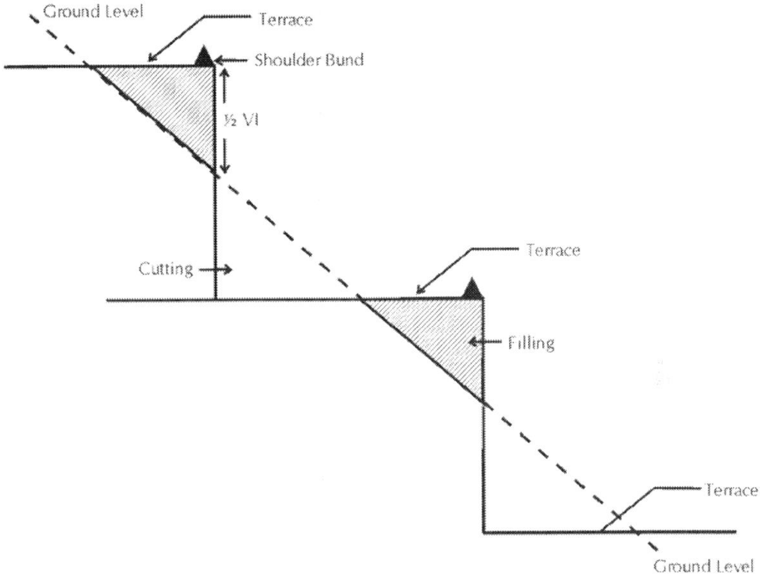

Fig. 4. Design of bench terrace

Terracing is one of the mechanical methods of controlling erosion by converting slopy land into strips with tolerable slope (varying according to soil type and rainfall) called terraces and provided with V- shaped drainage channels and shoulder bunds to take away runoff water. By terracing, the whole length of slope was broken into strips or steps so as to reduce the degree of slope to minimize soil erosion (by run-off) and help rain-water conservation in situ. However, as already stated, there is limitation of land-slope upto which it can be practiced. Ayers (1936) opined that the first step in any rational solution was to restrict open cultivated land to slopes and conditions where erosion can be held within permissible limits. The land capability classification rules out any land steeper than 12 percent as unsuitable for cultivation. This assumption was realistic in countries like United States and wherever flexibility existed in the choice of land. But in most countries farmers were compelled to cultivate even steeper land. According to Sharda *et al* (2006) terracing was generally recommended upto 33 percent land slope, although in Himalayan region it was being practiced upto 50 percent land–slope. Thus, terracing probably remained the best means of soil conservation on steep land if it had to be cultivated. Also, proper care must be taken in consideration of the site-specific agro-climatic and ecological conditions and the technology must be adopted to the socio-cultural and economic environment. The benefits of terracing are most easily recognised in areas of low rainfall because of water conservation (Mangrulkar 1970). The land area under crops on steep land increased on terracing-upto 25 percent on 30 percent land-slope. The terraces had to be well designed and constructed at regular intervals. The vertical interval (V.I.) was the elevation difference between the two succeeding terraces. Thus V.I. of the terrace determined the spacing between two terraces. It can be calculated using the simple formula:

$$V.I. = \frac{WS}{100 - S}$$

where , V.I. = Vertical interval (m)

W = Width of terrace (m)

S = Slope of land (percent)

5.3.5.1. Grade and length of terrace: Grade and length of terrace were to be decided according to soil type. On light alluvial soils, the longitudinal grade should be less than one percent to drain out excess water from terraces without damaging these. In case of inward sloping benches a slope of 1 to 2 percent was adequate. Also, a toe drain was provided at the lower end. On heavy soils (clay and clay loam) a longitudinal slope of 1.0 to 2.0 percent was desirable to avoid water stagnation on terraces. On level terraces for paddy, a shoulder bund along the outer edge of the terrace was provided so that run-off/irrigation water did not damage the terrace. The length of terrace was usually limited to 100m for operational efficiency and to prevent erosion. The excess run-off from terraces/drainage channels was disposed into stabilized water-way.

5.3.5.2. Width of terrace: The width of bench terraces was a function of both soil depth and land slope. On deep soil, the slope governed the width of terrace, and it could vary from 2 to 6 meters depending upon situation. Greater the slope, less was the width of terrace. Also, method of cultivation and the farming operations to be practiced (for a crop) and the source of power available for these operations had to be kept in mind while deciding the width of terrace, as bullocks and tractor needed a minimum width to turn. Construction of bench terraces on land with different slopes was illustrated in Figure 5 (A to C). As shown in Figure 5A, the terrace width of 6m was obtained on land with 10percent slope. The vertical interval (V.I.) was also less than one metre (0.7m). Thus the system was convenient for arable farming and could be easily stabilised and maintained. Also, both bullock power or tractor could be used. With land slope of 33 percent, the terrace width was reduced to 2.7 m while V.I. increased to 1.3 meters. The system became less stable, restricted in choice of crops and farm power, and more expensive to maintain. With land – slope of 50 percent, the width of terrace was reduced significantly to 1.4 meters, although the V.I. increased only slightly. The terraced land was more suited to silvipastoral/ hortipastoral system than arable farming. A simple design of terracing was suggested by Hurni (1986) according to which the vertical interval of one meter should be adopted for all land-slopes less than 15 percent. On steeper slopes, the V.I. should be 2.5 x soil depth.

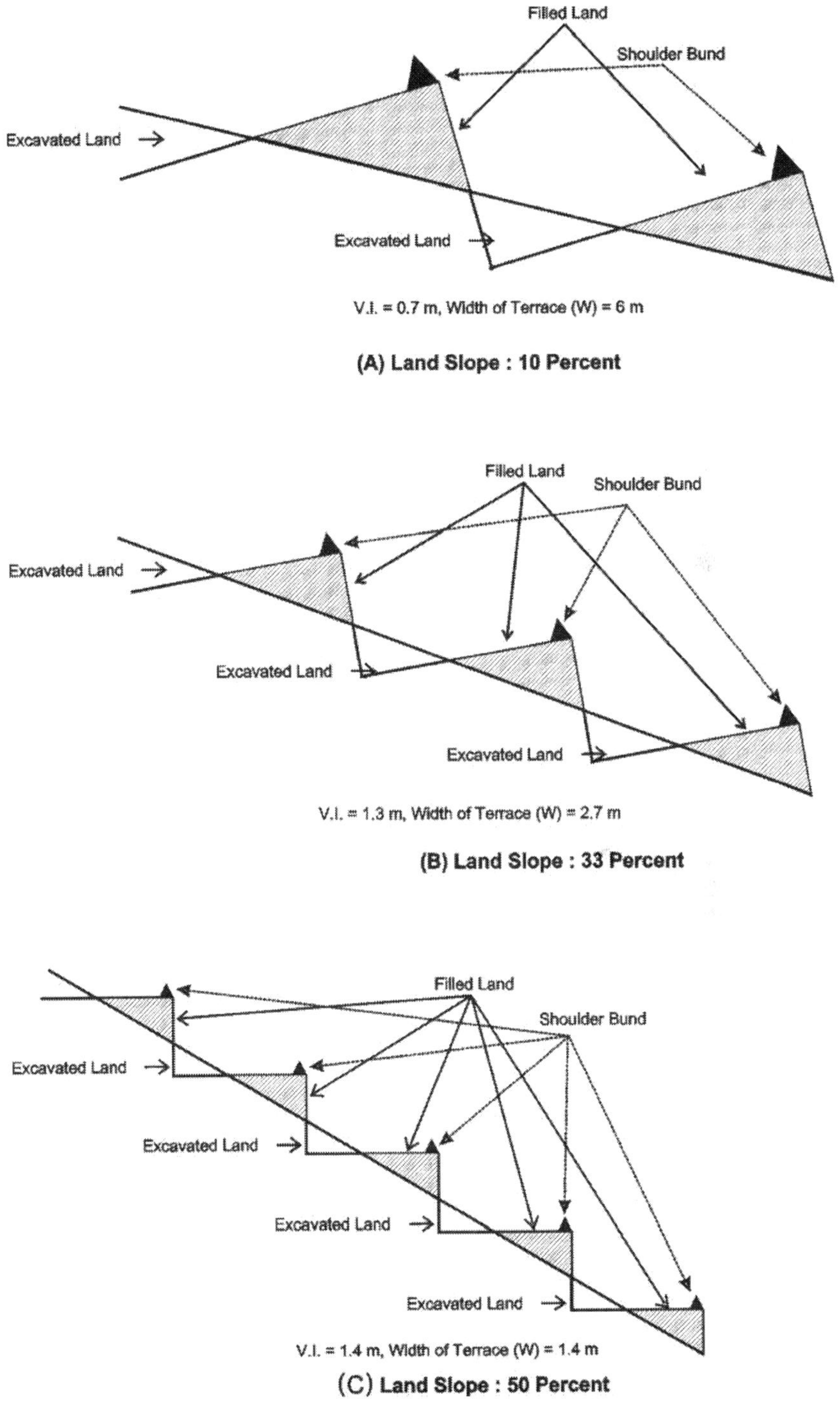

Fig. 5. Effect of slope on terrace width

Besides land-slope, depth of soil is another basic factor determining the terrace width. The possible width of the terrace without excavating into subsoil or bed-rock is a function of soil depth (Hudson, 1988). As shown in figures 6A and 6B, with same land slope, both the width and the vertical interval of terraces increased with soil depth.

The volume of earth moved increased with the width of the terrace:

$$C = W \times \frac{Hr}{8}$$

Where C is the area of cross-section (m^2), W is width of terrace (m) and Hr is the height of the riser(m) and is equal to vertical interval plus the change in elevation across the terrace if there is inward slope of the terrace (Hudson 1988).

5.3.6. Types of terrace

Depending upon the situation (soil type, rainfall, slope and degree of erosion, etc.) following type of terraces were constructed:

i. Bench terraces

ii. Broad based terraces: (a) graded terraces; (b) level terraces

iii. Zingg terraces

iv. Conservation bench terraces

5.3.6.1. Bench terraces: Bench terraces are series of platforms having vertical drops along the contours or on suitable grade lines across the general slope of the land. Depending upon the need, these might be either completely levelled or have a gradual slope. Bench terraces intended to maximize infiltration were level in both directions i.e. width wise and longitudinal direction, or had slight reverse slope. The terraces could have a slope inward or outward, depending upon the intensity and total annual rainfall. Also, appropriate drainage channels had to be constructed along the bench terrace to take away runoff water. These channels had to be designed based on at least 5 years' peak flows of rainfall and keeping a free board of 15 cm. On slopy land with shallow soil (hill slope), narrow bench terraces called step terraces were constructed and used for growing tree crops (tea or coffee). The earth moving in constructing these was much less compared to wide bench terraces.

Intermittent terraces or orchard terraces were constructed for growing larger tree crops such as rubber or fruit trees. Plate forms are short lengths of bench for a single tree or bush. The shape of plate form could be square, circular or hexagonal (Hudson 1988).

5.3.6.2. Broad based or Absorption type terrace: This type of terrace was constructed where water conservation was important. This consisted of a wide embankment (bund or ridge) constructed on the lower edge of the channel from where soil was excavated. The excavation of channel was done at suitable intervals on a falling contour with a suitable longitudinal grade. The broad based terraces were further subdivided into graded terrace and level terrace depending upon whether these were constructed on a grade or on dead contours. Level terraces were generally recommended and constructed on the contour in areas where the top soil is sufficiently permeable to prevent over-topping of the ridge and serious damage to crop, where as graded terraces are constructed on heavy soils with low infiltration rates.

*5.3.6.3. **Zingg terracing:*** Zingg terracing was suitable for arid and semi-arid areas in black soil region where land slope was 3 to 6 percent. The lower one third portion of the inter-bunded area was levelled to allow spread of run-off water on a wider area. The upland area was planted to dryland crops such as greengram, blackgram, maize and sorghum whereas lowland portion was covered by water intensive crop such as rice and soybean. In this system, the sloping land in the upper part was donor area whereas the lower part was the receiving area for the run-off and thus it resulted in maximum utilization of run-off water.

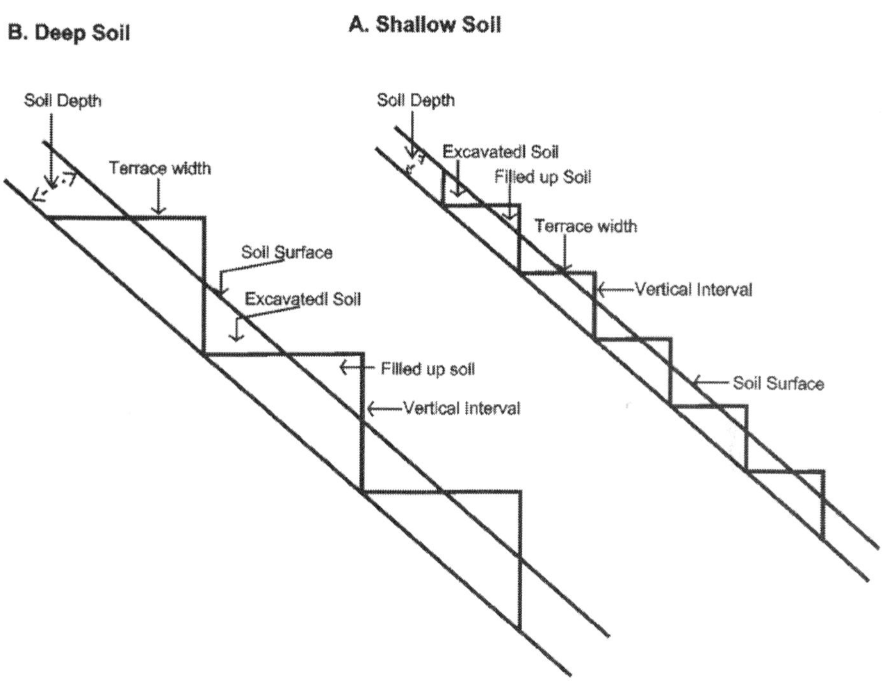

Fig. 6. Effect of soil depth on maximum width of terrace

*5.3.6.4. **Conservation bench terrace:*** This system was similar to zingg terracing but was recommended for high rainfall area where land slope was 3 to less than 6 percent and control of soil erosion in the donor area was necessary. The sloping (donor) area was accordingly kept under silivi-pastoral/horti-pastoral system and soybean or paddy was raised on level land (run-off receiving area). The donor area (upper slopy part) was left in natural state where as the lower slopy part was turned in to a level bench to receive run-off water and spread it uniformly over the whole area. The ratio of donor area to receiving area varied from 2:1 to 3:1 and level bench was spaced at vertical interval of 1 to

1.5m. The cross-section of shoulder bunds should be 0.35 to 0.50 m^2 for red and lateritic soils and 1.2 to 1.5 m^2 for black soils (Sharda et al 2006).

5.3.7. Slopy land (slope < 6 percent) with medium and deep soil

As already stated such land usually occurred downstream the steep land with shallow soils where the slope was less than 6 percent and soil depth increased up to one metre. Also, further down in the watershed, the slope decreased and soil depth increased continuously. If this land below hills (present on the periphery of watershed) was cultivated, it suffered more from high velocity run-off coming from upper area than its own run-off. The erosion problem on it could be taken care of by appropriate mechanical structures such as bunds, measures for safe disposal of surplus runoff and proper crop husbandry practices. Construction of bunds was the major and important mechanical measure for controlling the soil erosion on cultivated land but any sort of bunding like field bunding or property bunding, irrespective of soil type would not help. For proper utilization of water and its conservation, appropriate type of bunding depending upon soil type and climatic conditions was necessary to protect land against the damage by excessive runoff.

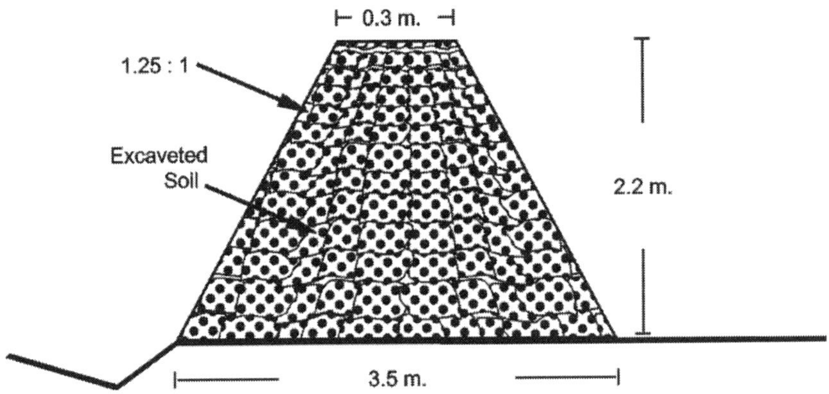

Fig. 7. Cross section of Water Diversion bund

5.3.7.1. Water diversion bund: The construction of a water diversion bund was the first line of defence in soil conservation work. Obviously, it was constructed between the steep land with shallow soil and the cultivated land. It diverted the foreign water running off hillocks or grasslands and then running through the cultivated fields. It was essentially a graded bund with a V-shaped drainage channel along it on a grade of 0.2 to 0.3 percent (Figure 7). The cross section of the bund depended upon the area of which the water was to be diverted and the rate of runoff from the area (Table 21). The bund should have a side slope of 1.25 to 2.0: 1.0. If the catchments included a bare hillock, the base width had to be increased by about 50 percent to take care of greater volume with high velocity of runoff to be diverted. Also, pitching with boulders was necessary at bends wherever these occurred to protect it from eroding away.

Table 21. Base–width of diversion bund

S. No.	Area of catchment (ha)	Base – width (m)
1	20	1.0
2	30	1.5
3	40	2.3
4	50	3.5

The water diversion bunds diverted water running off a hillock, grassland or any other type of land covered with vegetation, away from the cultivated fields, as bare soil in these fields was much more susceptible to erosion. Such bunds could lead water into a tank, on a grass strip, to a natural drain or into a grassed waterway constructed for this purpose, as the situation demanded.

5.3.7.2. Designing water diversion bund: For designing a water diversion bund, it was necessary to decide base width, top width, slope of the bund and its height. Knowing the base-width (Table 21) top-width (0.2 to 0.3m) and slope of the bund-side (1:1.25 to 2.0), the height of bund can be worked out and cross-section of the bund calculated. For example, if base width was 3.5 m, top-width 0.3m and slope of bund-side 1:1.25, the height (H) of the bund would be:

$$H = 1.75 \times 1.25 = 2.1875 \approx 2.2\text{m}$$

$$\text{Accordingly cross-section of the bund} = \frac{3.5 + 0.3}{2} \times 2.2\,\text{m}^2$$

$$= 4.18\ \text{m}^2$$

5.3.7.3. Contour bund: To take care of erosion in the cultivated area below the water diversion bund, the whole catchment needed to be divided into small segments through construction of contour bunds depending upon the soil type. A contour bund was one constructed on the points of the same elevation. Contour bunding was desirable on light soils which were permeable, i.e. had high rate of infiltration. A contour bund not only intercepted the water-flow but allowed it to walk rather than to run. It also gave sufficient time to the runoff water to in filter into light soil before running away from the catchment.

5.3.7.4. Graded bund: On land with slope greater than 3 percent but less than 6 percent, and with clayey texture, graded bunds were recommended in place of contour bunds. The system of level contour bunds held up the water and worsened the water-logging problem on clay soils. While contour bunds were a useful way of conserving water in low rainfall areas and on light soils, in high rainfall (above 700 mm) regions and particularly on Vertisols having low permeability, graded bunds were better in directing runoff water safely from crest to waterway. A graded bund comprised of a channel and a bund on suitable grade. The specifications of a graded bund regarding grade, spacing and cross-section were discussed below and the alignments of graded and contour bunds were illustrated in Figure 8.

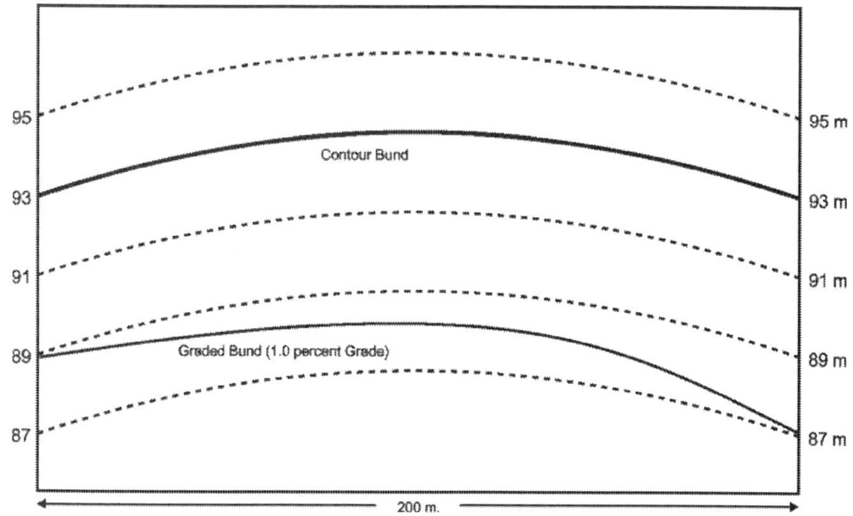

Fig. 8. Alignment of Contour and Graded bunds

5.3.8. Design of bunds

The prime factor governing the design of contour/graded bund was the depth of water to be impounded, which could be determined by the formula:

$$D^2 = FR + 16$$

where D was the depth of impounded water in feet, F was the vertical interval in feet between the two bunds and R was the total runoff in inches. According to Singh (2015):

$$H = \{(Re \times VI)/2\}^{1/2}$$

Where H was the depth of impounded water in metres, Re= excess runoff of 24 hrs in cm, VI= vertical interval in metres.

Other estimates required for the design are: (i) grade of bund, (ii) spacing of bunds and (iii) cross section of bund.

5.3.8.1. Grade of bunds: The grade in case of graded bunds should be as detailed below:

First 100 m length: no grade,

Second 100 m length: 0.1 percent grade,

Third 100 m length: 0.2 percent grade,

Fourth 100 m and beyond length: 0.3 percent grade.

The total length of the bund at a stretch should not exceed 500 m in low rainfall areas and 300 m in high rainfall areas.

5.3.8.2. Spacing of bunds: The information on spacing of bunds recommended for adoption on the cultivated land was presented in Table 22. The spacing may be as much as 150 m on land with a slope of 1 to 1.5 percent while only 40 m when the slope was 5 to 6 percent (Gadkary, 1966).

Table 22. Spacing of bunds for different land slopes

S. No.	Slope (percent)	Vertical Interval (m)	Horizontal interval (m)
1	0.8 – 1.5	1.3	110
2	1.5 – 2.0	1.5	85
3	2.0 – 3.0	1.7	70
4	3.0 – 4.0	1.8	50
5	4.0 – 5.0	2.0	45
6	5.0 – 6.0	2.2	40

5.3.8.3. Cross section of bunds: A trapezoidal cross-section was usually adopted for bund construction (Figure 7). The cross section of a graded bund (top width, bottom width and height) recommended for adopting on any type of land ultimately depended on land slope, depth of soil and the type of soil and the same were presented in Table 23.

Table 23. Cross section of bunds (m^2) for different situations

Land slope Percent	Cross section	Soil depth	Cross section	Soil type	Cross section
0 – 2	0.70	Shallow	0.45	Clay	0.65
2 - 3	0.75	Medium	0.53	Loam	1.00
3 – 6	0.85	Deep	0.60	Sandy	1.50

The cross-section of bund could vary widely from 0.45 to 1.50m^2 in different regions. To be safe, for any combination, maximum value should be accepted.

5.3.9. Ramps for crossing of bunds

Construction of a large number of bunds in any area or field made the access difficult. Ploughs and other heavy implements which were needed for field preparation were required to be lifted over the bunds. Also, there might be cart tracks to various holdings not necessarily located on the crest, and these might have to be maintained for access to fields by farmers. Therefore, an adequate number of ramps might have to be constructed wherever necessary. These may be earthen ramps or boulder plus *murrum* ramps with a slope of 1:3 to 1:10.

5.3.10. Sur-plussing arrangement

To dispose off safely the excess water that fell on the inter-bund area and ran off along the contour/graded bund into a stabilized wash, gully or a grassed waterway, an outlet or a waste weir was constructed at the junction of graded bund and waterway, depending upon the difference in the elevations of waterway bed and the base of graded bund. Besides this, the catchment area and the maximum intensity of the rainfall had to be considered while designing the waste weir.

5.3.11. Different types of waste weirs

Depending upon the specific site, one of the following types of waste weirs could be constructed:

i. clear over-fall stone weir

ii. a ramp-cum-waste weir

iii. a grass outlet

iv. a pipe outlet

5.3.11.1. Clear over-fall stone weir: A clear overfall stone weir should be provided on the bank of the waterway where a graded bund joined the waterway if the difference in the bed levels of the base of graded bund and the waterway-bed was greater than 80 cm. Also, its crest wall top should be 20 to 25 cm. above the level of the bund base. Such a weir might comprise of a masonry wall or a gabion of designed length with pitching on both the ends of the weir.

5.3.11.2. Ramp-cum-waste weir: A ramp-cum waste weir was necessary when there was a difference of 40 to 80 cm. between the bed levels of waterway and the foot of the graded bund and when the catchment of the bund was more than 10 ha. This was made by boulders arranged in the form of a slopy base, and pitching of the bank of the waterway on both the sides on the inlet. The crest wall of the weir should be 20 to 25 cm. above the level of the foot of the bund.

5.3.11.3. Grass outlet: A grass outlet could substitute a channel weir or ramp-cum-waste weir if the catchment of the graded bund (i.e. inter-bunded area) was less than 5 ha and if the grass could be established quickly or if the grass was already established at the point where surplus water was to be let into the waterway. The crest of such a weir should be 20 to 25 cm. above the bed level of the waterway.

5.3.11.4. Pipe outlet: A pipe outlet comprised of a piece of pipe discharging the surplus water through it. The design consisted of a hume pipe of required diameter with one "Well" on the upstream side. The diameter of the pipe depended upon the inter-bund area of which the surplus water was to be let into waterway, i.e. diameter of the hume pipe recommended for the outlet, varied with the area of its catchment e.g. 15cm for a catchment of 3 ha, 25cm for 5 ha and 30 cm for 30 ha.

5.3.12. Length of the crest wall of waste-weirs

The length of the crest-wall of the waste-weir was determined by the size of the catchment and 5 year peak runoff of the area. A flow of 30 cm of water over the crest-wall for small catchment (upto 30 ha) and of 60 cm to one metre for large catchments (greater than 30 ha and upto 100 ha) was assumed. The length of the crest-wall recommended could be calculated accordingly after calculating the quantity of run-off available.

5.4. STABILIZATION OF WASHES

Formation of washes through the cultivated fields by runoff water was the first stage of gully-development and needed to be stabilized so that these did not grow up with time as a result of continuous bed erosion. The stabilization of washes could best be done by leaving the strip of land, over which water flows, unploughed so that grass got established gradually. If establishment of natural grass took too long, suitable grass species had to be planted, if the wash was getting deeper and wider every year due to bed erosion at first and then falling away of sides. For a shallow wash (depth upto 50 cm), a rectangular weir type drop structure of loose boulders at an appropriate site with a reverse filter (consisting of pebbles, gravel and sand layers) upstream the structure, was sufficient.

If the wash was deeper than 50 cm or if the velocity of run-off was greater than 1.5 metre/second, an appropriate gabion structure would be necessary for its stabilization. A gabion was made by packing boulders into boxes made of nets woven from galvanised iron (G.I.) wire. The foundation was dug just like a masonry structure and a reverse filter made at the bottom of the foundation pit. Above the filter, the boxes of G.I. wire nets were placed and packed with smaller boulders at bottom and comparatively large ones at the top. Structures of various sizes and shapes could be made by placing a number of boxes together and at the top of one another as required. Gabions are more effective than masonry structures of the same design because these were porous and allowed runoff water to pass through while retaining the silt upstream. These structures were particularly valuable for black soils on which masonry structures cracked and gave way due to shrinking and swelling property of these soils.

5.5. GULLY CONTROL AND RECLAMATION

Since gully formation was caused by concentrated run-off, farming practices which reduced quantity, velocity and concentration of run-off and increased infiltration or encouraged uniform infiltration served as preventive measures against gullying because only a minimal amount of rainfall got concentrated and run off. If these practices were not adopted, with concentration of runoff water from a number of washes, the main wash had a tendency to develop into a gully unless stabilized. The gully again went on deepening and widening with time due to bed erosion at first and then falling off banks. The process did not stop here and the side gullies started developing from the secondary washes. Thus, a net-work of gullies was formed wasting a lot of fertile land, unless the main gully was controlled and developed into a stable grassed waterway. Gullies were formed on cultivated level land with deep soils of different types, viz. alluvial soil, black soil and mixed red and black soil, when concentrated run-off water from upper parts of the watershed ran even on level land in the lower parts of the watershed. The run-off might get concentrated either by combination of run-off from many washes or when run-off was guided along a water diversion bund, peripheral bund, and graded/ contour bund.

Shallow gullies (less than 30 cm deep) could be economically stabilized by vegetation and used as fuel and fodder reserves. This area, which had been lost to agriculture, could thus be profitably utilized. Where vegetation could not be established quickly or where a medium gully (depth 0.3-1.0 m) was involved, the same could be plugged by small loose boulder structures, starting at the point where it originated. The height of this stone plug depended upon the elevation of the spot of its construction and that of the starting point of the gully. If necessary, brush type structure could also be constructed instead of loose boulder plug.

To reclaim a deep gully (depth greater than one meter) stabilization of both the bed and the sides of the gully was necessary. For this, construction of drop structures in the gully at appropriate points and planting grass at the bed and on the sides of the gully were necessary measures. The drop structures made of gabions of appropriate dimensions had been found to be very effective for the reclamation of gullies (Figure 9), particularly on black soils. (Indo-UKDFP 1980a).

5.6. RAVINE RECLAMATION

As already stated, on highly erodible alluvial soil, a net-work of deep gullies (ravines) was formed resulting in complete loss of fertile land. The experience of gully-control on black soils could not be transferred as such to alluvial soils for ravine-reclamation because Chambal Valley soil had much higher dispersion coefficient and pitching of earthen bunds/banks was not successful unlike black soils. Also, the problem of caving under the structures was serious. For ravine control and reclamation, management of rain-water on watershed basis was the only appropriate approach. Rain-water had to be managed starting from the ridge through level-land (table land) on the upper part of the watershed where ravines might not be a problem. Main gullies draining the table land were to be identified and stabilized as grassed waterways. The inter waterways area should be divided in to small segments by construction of contour bunds across the slope, running from one waterway to another and draining the segment of land into the waterway. The inter-waterway land with shallow and medium ravines could be reclaimed just by alternate land-use, i.e. by establishing agri-horticultural or silvi-/horti-pastoral systems in place of traditional crops. Reclamation and control of deep ravines, particularly those like Chambal ravines, was very tedious job. Afforestation of ravines was often recommended but it was almost impossible to plant on very steep and inaccessible sides of very deep ravines (up to 25 meters deep). Afforestation through aerial seeding undertaken in the World Bank Project was an utter failure. Even otherwise, afforestation as a single measure could neither reclaim nor control the formation of ravines. Alternatively, the problem has to be tackled on watershed basis, and the first step was to indentify the main ravine (gully) of watershed/sub-watershed and stabilise it with suitable mechanical measures. Similarly suitable structures at appropriate sites on side-gullies have to be constructed before the angular tops suitable size for plantation of trees for fuel, fodder and fruits. The structures would help to retain the soil (excavated) in the bottom of ravines. To stabilize steep slopes left after limited levelling of peaks of ravines, suitable grasses and plants like munj (*Sacharum munja*), agave, etc. should be planted on slopy land. Thus, construction of structures, limited land levelling and plantation of trees for fuel, fodder and fruits, medicinal plants, bamboos and various grasses, etc. were the necessary steps to be taken to reclaim and control ravines and utilize the ravineous land gainfully and on sustainable basis (Verma and Singh 2009).

Fig. 9A. Gabion

Fig. 9B. Gabion in Rainy Season

5.7. GABION STRUCTURES FOR STABILIZATION OF WATERWAYS AND GULLY CONTROL

Gully erosion is quite severe throughout the Central India, particularly where concentrated run-off flows fast through cultivated land and along both the banks of rivers. The type and magnitude of gully-formation depends on the slope of land. On slopy land (slope greater than 3 per cent) gullying is in finger-shaped pattern that is a number of small gullies originate as a result of washes and converge into a single big gully on comparatively flat land. On flat land (slope less than 1%), the main gully is zig-zag resulting from meandering stream and is joined by a few side gullies. All gullies march forward towards the ridge and also become deeper and wider with time. At first, the gully-bed gets eroded and then side banks become instable and fall off. Thus lot of land gets gullied every year. This requires stabilization and reclamation of gullies in order to prevent further deterioration, prior to the development of in-field works for safe disposal of run-off water. For the stabilisation of a deep gully, a huge gabion (Fig. 10) must be constructed.

Fig. 10. Huge Gabion

5.7.1. Why gabion structures?

Planting of grass on the waterway-bed or on the bed of gully to stabilize it, is difficult if slope is greater than 1 percent. Structures of loose rock will, however, adjust to such movements, but moving water and errant cattle can easily dislodge the rocks, and eventually partly or completely demolish the structures. Also, such structures are successful under field conditions only for a fall of 50 to 60 cm. For a bigger fall (up to 2m), the use of semi-flexible structures of lose boulders properly packed into prefabricated G.I. wire boxes known as 'gabions' offer a possible alternative to either of the above methods. Other reasons for the choice of gabion structures are:

1. The use of masonry structures (using bricks/dressed boulders and cement) for such works in areas characterized by Vertisols (which swell and shrink on wetting and drying) has not proved successful, as these crack and collapse following soil movement. Galion structures are flexible and porous and consequently adjust to movement of soil. Being porous and having reverse filter, these retain silt and sand particles upstream and allow only water (run-off) to flow through.
2. Unlike masonry structures, water and cement are not required to construct the gabions and this is a big advantage in rainfed areas.
3. If any gabion structure is defective in construction or gets broken, it can be reconstructed using the same material.
4. If net boxes of required size are not available as prefabricated ones, the same can be fabricated at the site by local labourers/farmers, using just G.I. wire.

5.7.2. Gabion structure most appropriate for Vertisols

Gabions of different sizes (mini-gabions to huge gabions) were tried successfully in Indo-UK Dry Farming Project area to stabilize waterways and for gully-control as early as 1976. Gabion of different sizes were also tried in the same watershed area to stabilize big *nalas*. The details of construction of such gabions are given below (Shaxson 1977).

5.7.3. Selection of site

The main purpose being to stabilize the bed level of a gully, a primary location of the structure is at the foot of an unstable and eroding section of gully, where it runs on to a stable gradient. The height of notch will determine how far back up the eroded section silt will be deposited. If this does not throw back as far as, or beyond, the head of the unstable section, the notch height has to be increased, or a further structure should be constructed just above the furthest point of throw back of the previous structure. This produces a 'staircase' effect, with section of gully bed in stable condition falling to lower elevations over the structures. Also, if a side gully joins the main gully, the junction point provides a better site for gabion structure, as it will stabilize both the side and main gully.

5.7.4. Design

Following the measures of depth and width of the gully/waterway at the selected site, the structure is designed in the light of required notch width, notch height, freeboard available and the extent of apron. The reference level is to be that which has been chosen at or below gully bed level at the downstream edge of the apron. Figure 11 shows the different components of a gabion structure. The components are described as follows:

5.7.4.1. Headwall notch: The notch must be at least as wide as gully so as not to constrict the flow. The notch width varies according to (i) value of 'Q' (ii) whether it is designed with or without over fall, and (iii) depth of flow. The length of notch (in metres) is calculated for different values of 'Q' and for different flow-depths (h). Flow over the notch is calculated under free flow conditions as follows:

$$Q = C \times L \times H^{3/2}$$

Where Q = designed discharge (m^3/sec.)

C = constant 1.79

L = length of notch (m) and

H = height of flow over notch (m)

5.7.4.2. Wing walls: Notch height can be achieved by available sizes of gabion box, which (using the 'least-waste' specifications) are units or combinations of one metre and 0.4 m in height. Alternatively, boxes of required size can be prepared from G.I. wire nets, knitted at the site using G.I. wire.

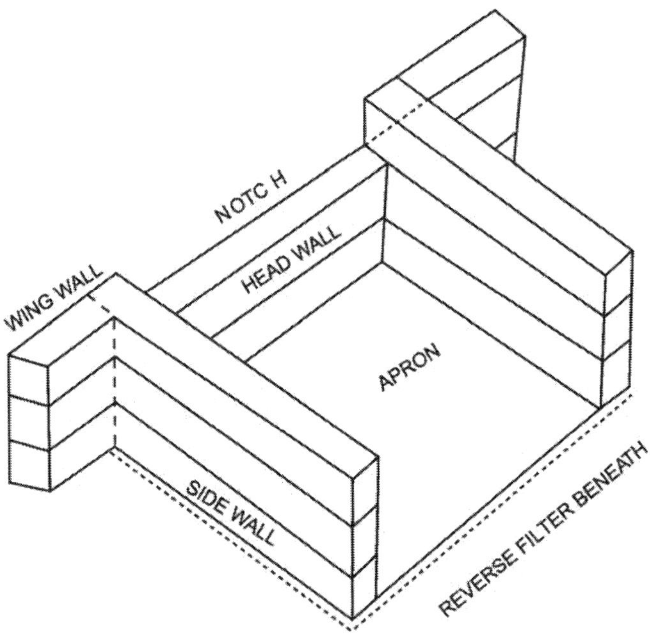

Fig. 11. Different components of a gabion.

These must be extended upto 2 metres into the gully bank on either side, and must be rendered impermeable by any appropriate means using an impermeable sheet or rammed damp clay. Only where a gully is less than 2 metres wide and 0.5 metre deep, this distance can be reduced to 1 metre and that too only after careful appraisal of the risk of water cutting round the end. Also, in Vertisols, 1m extension of wing wall into gully bank on either side is sufficient.

5.7.4.3. Side walls: These must extend at the lowest elevation not less than 3 metres downstream of the headwall. Where overall height at the notch is more than 1 metre, the side walls should extend an extra 3 metres for every 1 metre extra height. They will be built either vertically of 1 metre high boxes, or slopped up the gully side (as appropriate) using the 0.4 metre thick units used for mattress. The side walls will be cut into the gully walls so as not to constrict the gully width, since the notch itself must extend the full width of the gully.

5.7.4.4. Mattress/Apron: This will provide a bed of 0.4 m thickness upon which the water will fall from the notch, and will extend to at least twice the notch height downstream. Also, it will extend beneath the entire structure including headwall, wing walls and side walls.

5.7.4.5. Reverse filter: A filter of permeable non-destructible materials, viz. coarse sand and small stones (*gitti*) placed in order is called "reverse filter". The coarse sand and small stone layers are laid before the gabion wall so that water passes through sand before it passes through small stones and ultimately the gabion wall. The reverse filter is required below and around the entire structure (except upstream of wing walls) to permit the passes of water through the structure, but to prevent passage of fine soil particles.

5.7.4.6 Freeboard: A minimum freeboard of 0.3 m must be added above the planned maximum depth of flow over the notch.

5.7.5. Excavation

The reference level for survey work and construction should be fixed at the downstream of the apron such that it is not disturbed during field-works. This must be precisely to the size, with walls vertical, corners square, and all surfaces cut evenly down to the planned levels. The lowest level to which the excavation will be done is the reference level (reduced level of gully-bed) minus 0.6m (mattress 0.4m + reverse filter 0.2 m). Excavation for head-wall extension at higher elevation marked only after this stage is completed.

5.7.6. Materials required

Various materials required for construction of gabion structures are:

1. Prefabricated net boxes of size 1m x 1m x 1m and 1m x 1m x 0.40m
2. The net boxes can be made of G.I. wire net or chain link or crimped net
3. If prefabricated net boxes are not available, only G.I. wire is needed. It can be woven into net at first and then boxes can be made of this net.
4. Stones or boulders of different sizes.
5. G.I. wire for tying together the boxes or for tension within the box.
6. Sand and small stones (*gitti*) for reverse filter.

5.7.7. Quality of wire for weaving gabion boxes

The quality of suitable wire for weaving gabion boxes depends on its gauge, tensile strength, resistance to bending and extent of galvanization. Based on experience of Indo-UK Dry Farming Project, Indore, where gabion boxes were used under such conditions that on problems were posed by abrasive rocks and debris in run-off water, the following specifications of wire were appropriate (Shaxson, 1977).

1. Gauge : 10 (6.5kg/100m)
2. Tensile strength : medium (40kg/mm^2)
3. Rust-proofin : Double galvanization
(Zinc coating 70g/m^2, coating done after drawing wire)

5.7.8. Construction

The prefabricated boxes are brought onto site and placed in position, one layer at a time. They are then wired together and tensioned. Stones (boulders) of the necessary sizes must be carefully placed and packed into the boxes, and the sides wired across one to the other at the mid-level of filling in case of 0.4m high boxes for mattresses and at the 1/3rd and 2/3rd levels of filling in case of 1 m high boxes.

5.7.9. Precautions for construction of good gabions

Following precaution are necessary to construct a good gabion:

a. Tightness of the mesh and wiring is essential at all times.
b. Special care must be taken to ensure that all joints between side panels (crimped net) and the diaphragms ends, tops and bottoms (of chain link mesh) are adequate in number and security. This must be checked, and if necessary, remedied before filling begins.
c. Each level of gabion unit is to be completed before the next level is started.
d. Packing of boulders in GI wire boxes should be done such that smaller bounders are in the lower part of the gabion and bigger ones in the upper part.

5.8. GRASSED WATERWAYS

As already stated, gullies formed by concentrated run-off had to be stabilized and reclaimed. Also, to encourage overall drainage and to lead water diverted by storm drains or water diversion bunds to natural streams, grassed waterways had to be constructed at suitable sites. The dimensions of waterway (height of banks and depth) was calculated based upon the rate of runoff, its total catchment area and the value of permissible velocity of water (in the waterway) which depended on the nature of bed e.g. 0.5 m/second for bare sandy soil, 1.0 m/second for bare soil, 1.4 m/second for grassed soil and murrum, and 1.7 m/second for hard rock. If the bed of waterway was too steep, it got corroded when the velocity of water exceeded these limits.

If necessary, beds of waterways could also be stabilized by construction of appropriate gabion structures of appropriate size at proper points. Also banks at the bend of a waterway had to be protected by construction of spurs or diversion-walls made of gabion boxes. A waterway with proper bed gradient got stabilized by natural vegetation.

The rate of run-off varied according to type of catchment and it could be reduced by treating it suitably (afforestatation of bare hillock, establishment of grass by preventing overgrazing etc.). Also, the load on a waterway could be reduced by either diverting the water safely (without causing problems elsewhere) or creating a water reservoir above the waterway.

5.8.1. Designing Waterways

After knowing Q, the cross section of waterway 'A' could be calculated knowing the permissible velocity of water to be taken away by the waterway. From 'A' values, channel dimensions could be worked out for a particular shape, e.g. V-shaped, parabolic or trapezoidal one. Some examples worked out as illustrations (Kale, 1977) were presented as follows.

5.8.2. Basic data for designing waterway

Suppose the catchment consisted of partly hillock and partly grassland and other data were as given below:

Area of catchment	-	48 ha
Expected maximum runoff	-	0.25 m^3 /ha /sec
Permissible Velocity (V)	-	1.5 m/ sec
Side slopes	-	1 : 1.5
n (Manning's roughness coefficient)	-	0.03
Bed slope (S) - 1 in 300 (0.33 percent)		

5.8.3. Designing V- shaped waterway

Q (Total run-off) = 0. 25 x 48 = 12 m^3/sec =A x V

where A=area of cross-section

V= velocity of run-off

$\therefore A = 12 \div 1.5 = 8 \ m^2$

Assume a waterway with a side slope of 1:8

A = ½ base (t) x altitude (h) where, t =16

$\therefore 8 = \frac{1}{2} (16)h$

$\therefore h = 1$

For checking the design, Manning's formula to be used:

$V = 1/n \times R^{2/3} \times S^{1/2}$

where n (Manning's coefficient)= 0.03

R = hydraulic radius,

S = hydraulic slope

$$P = \text{wetted perimeter} = t + \frac{8h^2}{3t} = 16 + \frac{8}{48} = 16.16$$

$R = A/p = 8 \div 16.16 = 0.495$ m

The velocity (V) $= 1/n\ R^{2/3}\ S^{1/2}$

$= 1/0.03\ (0.495)^{2/3}\ \ (1/300)^{1/2}$

= 1.21 m/sec which was within the permissible range

The final shape of this type of waterway was shown in Figure 12.

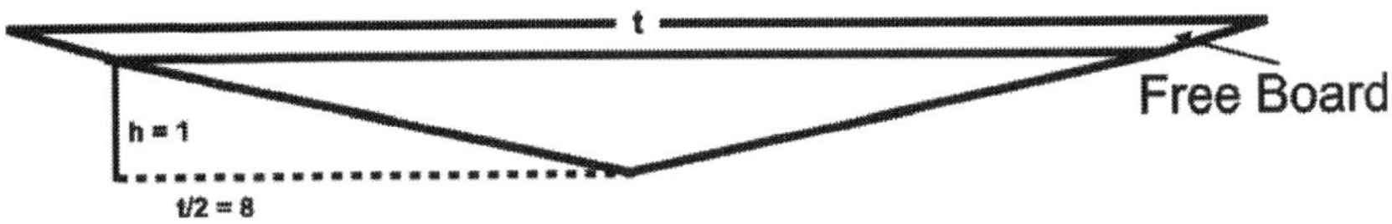

Fig. 12. V-shaped waterway

5.8.4. Designing trapezoidal waterway

$Q = 0.25 \times 48 = 12\ m^3/sec = AV$ where V= permissible velocity

A= cross section of water way

$A = 12 \div 1.5 = 8\ m^2$

Assume waterway of following dimensions:

B = base of trapezoidal waterway

B+3D = upper side of waterway

D = depth of waterway

$$A = \frac{B + B + 3D}{2} \times D = BD + 1.5\,D^2$$

$A = 8D + 1.5\ D2 = 9.5\ m^2$ when B = 8 m and D= 1m

If D was assumed to be 0.86 metre,

$A = 8 \times 0.86 + 1.5\ (0.86)^2$

$= 7.989\ m^2$ which was close to $8\ m^2$

For checking the design, Manning's formula to be used:

R = A/p

where R = hydraulic radius

P = wetted parameter

= 8+1.55+1.55=11.1

The velocity $= 1/n\ (R)^{2/3} \times S^{1/2}$

$= 1/0.03\ (0.72)2/3 \times (1/300)^{1/2}$

= 1.549 m/sec. This was within permissible range

Final shape of this waterway was shown in Figure 13. Also, since trapezoidal waterways are most common, their dimensions and cross-sections for different sizes of catchments (drainage areas) are shown in Figure 14.

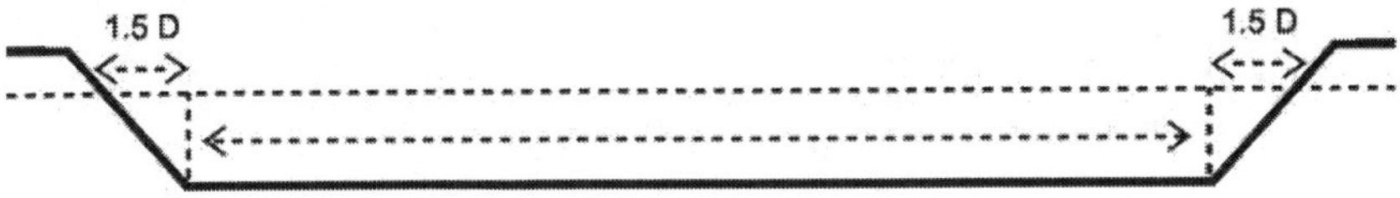

Fig. 13 A. Trapezoidal waterway section

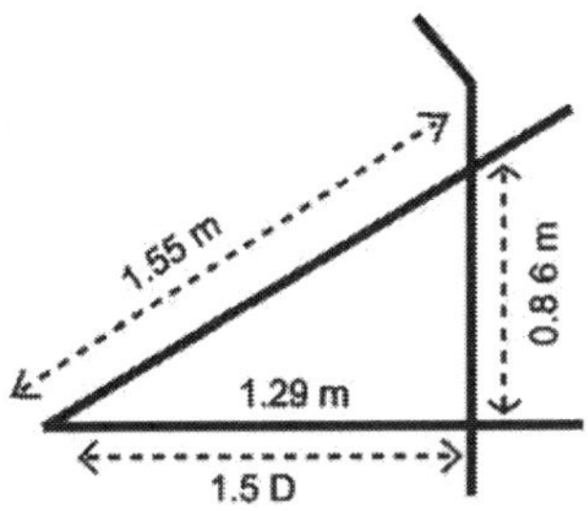

Fig. 13 B. Slanting surface of waterway

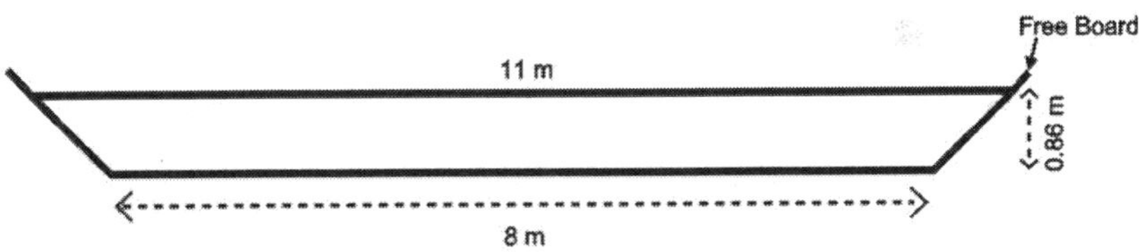

Fig. 13 C. Final shape of a Trapezoidal waterway

Fig. 13. A to C Trapezoidal Waterway

5.8.5. Designing parabolic shape waterway

As already calculated, A = 8m^2

Assume waterway of following dimensions:

Assuming D = 1 m, A = 2/3 T

: T= 12 m

For checking the design

$$P = T + \frac{8D^2}{3T} = 12.22\text{m}$$

R = A/P or 8 /12.22 = 0.6546

The velocity = 1/n (R) 2/3 (S) ½

= 1/0.03 (0.6546)$^{2/3}$ (1/300)$^{1/2}$

= 1/0.03 x 0.75 x 1/17.32

= 1.44 m/sec. This was within permissible range

Final shape of this waterway was shown in Figure 15.

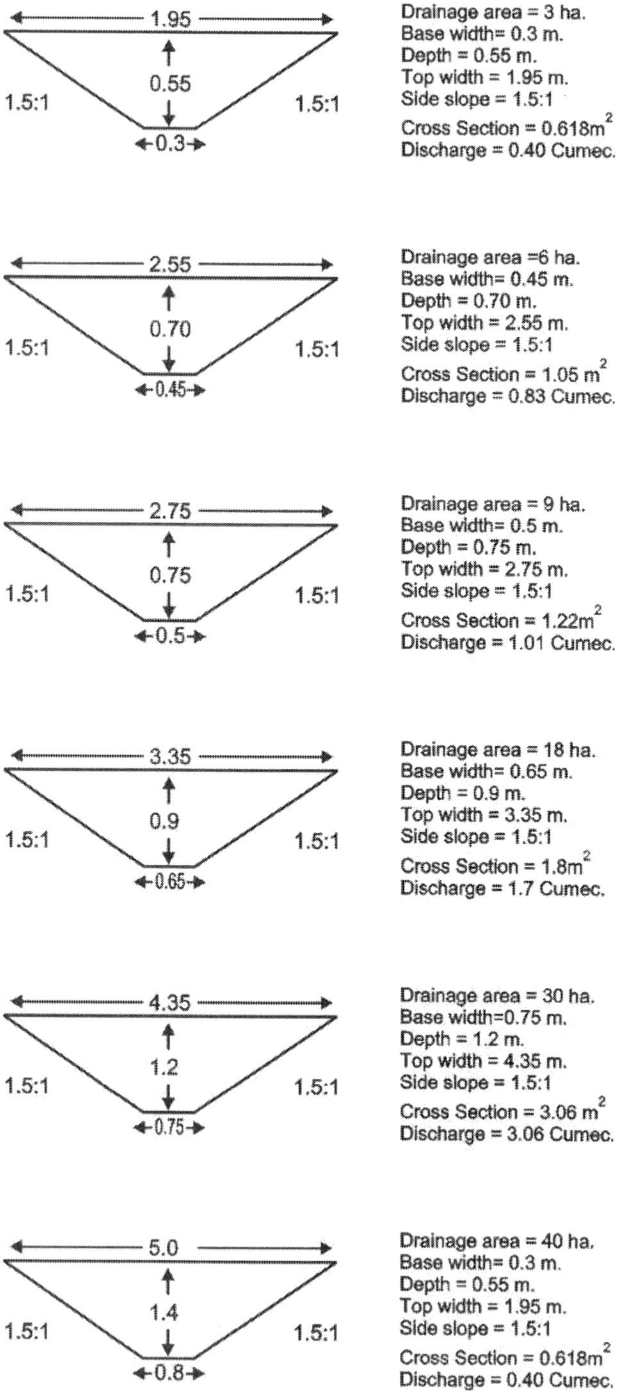

Fig. 14. Dimensions of a Trapezoidal Waterway for different sizes of drainage areas

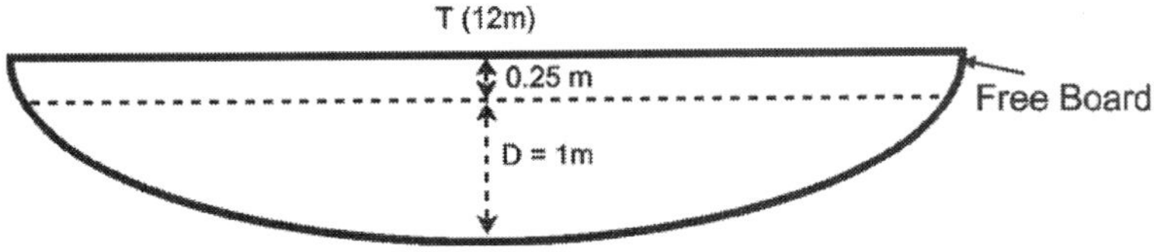

Fig. 15. Final shape of a Parabolic Waterway

5.9. MODERATELY SLOPY LAND WITH MEDIUM AND DEEP SOIL

If the cultivated land has a moderate slope (1 to 3 percent) the erosion problem was not serious particularly when average annual rainfall was less than 1000 mm. Further, if the steep and slopy land situated above the cultivated land had been treated appropriately and stabilized through measures already suggested (establishment of silvi-pastoral/horti-pastoral systems and construction of appropriate structures), the run-off water coming from above could not cause so serious problems as to make construction of mechanical structures such as water diversion bund and contour/graded bunds necessary. The mechanical structures tend to be expensive though quick enough to convince the farmers of good results, if the recommended works are executed timely and scientifically. However, conservation of soil and water through appropriate vegetative cover is being emphasized presently, being less expensive and easy to execute as compared to earth works.

5.10. Soil Conservation by vegetative cover

Conservation of soil and water through appropriate vegetative cover has been emphasized rightly, being less expensive and easy to execute as compared to earth works. As already stated establishment of an appropriate plant cover was not only extremely cheap (less than tenth to one hundredth the cost of earth work) and quite effective soil conservation measure (Shaxson 1980) but it also led to best utilization of land and water resources. Also, the farmers could do the work themselves. Plantation of trees for fuel, fodder and fruits and grasses on steep hill slopes and non-arable slopy land resulted in not only increased biomass production from hillocks and steep land but also reduced the rate of run-off and consequent soil erosion. Similarly for arable land, soybean had proved to be a good cover-crop, although a well nourished sorghum or maize crop also provided a good canopy. That is why growing a cover crop instead of leaving the land fallow in monsoon season has been strongly recommended. Also, if slope of the land was more than 6 percent and the soil was too shallow to do bench terracing, the land should not be used for arable farming but should preferably be diverted to silvipastoral or hortipastoral systems. Various improved grasses which could be grown on slopy land to arrest soil-erosion were Dinanath grass (*Pannisetum pedicellatum*), Hybrid Napier (*Pennisetum purpureum*), Guinea grass (*Panicum mascimum*), Blue pannic (*Panicum antidotale*), and Para grass (*Brachiasia mutica*). Besides providing an effective vegetative cover, grasses increase the infiltration rate of water into soil and consequently reduce the runoff. The practice of growing legumes with grasses resulted in not only more nutritious fodder but also minimized the use of nitrogenous fertilizer to pastures. A grassland after establishment had to be properly managed if good fodder, both in quantity and quality, was to be harvested from it, and if it was to protect soil against erosion by providing a continuous and effective cover. The role of vegetative hedges, contour/graded farming, strip cropping, mixed cropping and allay farming has already been emphasized under agronomical practices to make vegetative cover more effective tool of soil conservation.

5.10.1. Limitations of vegetative cover

The vegetative cover could not be prescribed as panacea for all ills of soil erosion. There was a limit to which vegetative cover could protect the soil from soil erosion. Also, it was not so easy to establish the vegetative cover on farmers' fields as under controlled conditions prevailing on research stations and state farms. Further, while the role of vegetative cover in preventing soil erosion to a certain extent was established, it could not reclaim the eroded soil to its original status. Again, protection against soil erosion by vegetative cover was limited to splash erosion, sheet erosion and initial rill-formation only. Once gullies and subgullies started developing, vegetative cover could not control their advance into fertile land. Beyond 3 percent land-slope in semi-arid, subhumid and humid regions, serious problems of soil erosion were observed. Even in dense forests, deep gullies and gorges were formed. In Chambal valley, even under natural conditions with natural vegetative cover existing, deep ravines were developing, getting deeper and wider with time and expanding their net-work, because soil was too fragile and under such conditions construction of appropriate structures was the only alternative. Of course, the role of vegetative cover was important to stabilize some of these structures at least in beginning, e. g. earthen bunds, earthen gully plugs, etc. The mechanical structures tended to be expensive but were effective to quickly convince the farmers of good results, provided works were executed timely and scientifically.

5.10.2. Vegetative hedges

As an alternative to earth-work (construction of water diversion bund, graded/ contour bunds etc.) establishment of vegetative hedges on contour/grade was emphasized. A series of vegetative hedges across the slope were established, maintaining the same spacing between these as recommended for graded/ contour bunds presented already in Table 22. As regards selection of plant species for vegetative hedges, vetiver grass (*Vetiveria zizanioides*) or *khus* has been recommended due to its prolific root system and hedge forming tendency. Other suitable grasses/plants e.g. *Saccharum spontaneum* (*kans*), *Cenchrus ciliaris* (*anjan*), *Iseliema laxum* (*musel*), *Cenchrus setigerus* (*pila anjan*), *Chrysopogon fulvus* (*Dain grass*), *Sehima nervosum* (*ponia*), *Dicanthium annulatum* (*marvel*), *Pennisetum pedicellatum* (*dinanath*), *Thalipogon aligans* (*tor*), *Cymbopogon flexuosus* (*lemon grass*), *Thysanolaena maxima* (*broom grass*), and *Saccharum munja* (*munj*) can also be tried. Of these, *munj* was better as its above-ground part produced fibre and other useful material which could be harvested without disturbing the vegetative hedge formed by its leaves. Unlike *munj*, vetiver could not be popular as its useful product was its roots which could not be taken out without destroying the vegetative hedge.

5.10.3. Contour hedge

Contour hedge, if established well, led to in situ moisture conservation and built up terraces with time by holding the soil particles upstream. Further such hedges did not need any maintenance and lasted for years effectively (ICAR 1987). In high rainfall areas where erosion was serious even at three percent land-slope, the effectiveness of barriers could be improved by raising a miniature bund along the hedge on the downstream side. However, it was not an easy task to establish effective vegetative hedges, particularly under farmers' conditions where free grazing was a right of the graziers, if there was no crop in the fields. Also, it was difficult to save the vegetative hedges during hot summers due to lack of irrigation water.

5.11. AGRONOMICAL PRACTICES

Besides establishment of vegetative cover and vegetative hedges as a measure of soil and water conservation, there were a number of anti-erosion agronomical practices (i.e. conservation agronomy) which controlled soil erosion. The same are discussed in continuation.

5.11.1. Crop rotation

This refers to growing of different crops in regular succession, on the same field or the same piece of land. In order to control erosion, the crop rotation must be such that it includes an erosion-resistant crop during the monsoon season. In rabi, the crop should preferably be deep-rooted and simultaneously building up the organic matter. Also, inclusion of legumes and grasses should be considered more often as part of the crop rotation, not only to improve the soil by increasing its nitrogen and organic matter content but also for protecting the soil against the massive erosion.

5.11.2. Raising a cover crop

By raising cover crop was meant growing of a good crop on the land for providing a vegetative cover. The land should not be left fallow during rainy season, and the crop to be grown should preferably have a good canopy to protect the soil effectively. Those crops which had vigorous growth and a smothering canopy with high leaf area were better suited for this. During rainy season, cultivating cowpea, green gram, black gram and soybean crop helped in providing very good cover. Raising these crops instead of keeping the land fallow was therefore advised, particularly on land susceptible to water erosion.

5.11.3. Crop residues management

Crop residues of last year grown crop if managed properly play an important role in promoting conservation farming. If crop residues are left undisturbed on soil surface, these serve as stubble mulching. Stubble mulching was the technique of protecting the surface soil during monsoon season by stubbles or other residues of the crop grown in the previous season. If the land could not be cultivated in *kharif* due to some socio-economic or other reason, it was better not to follow the practice of burning the crop residues (leaves, stubbles, etc.) left by the last grown crop. That is, the stubbles and other crop-residues should not be burnt as was the practice presently. Retention of crop residues on soil surface moderates the soil temperature and conserves moisture. If the residues are incorporated into soil, they built up organic matter content of soil and accelerated the microbial activity in soil and did not lead to termite problem which could be serious if residues were retained on the field (Tomar 2012).

5.11.4. Minimum Tillage

Since tillage was a costly operation (requiring much energy) and deep tillage encouraged soil erosion, the tillage practices just adequate to control weeds and prepare a good seed-bed were to be adopted. Direct planting after ploughing without a number of other field operations to have a fine seed bed was generally termed as minimum tillage. This was advantageous because it not only reduced the number of tillage operations and thus the cost of cultivation but also reduced the soil erosion and maintained the soil structure. The soil loss and erosion after minimum tillage was reduced to less than one fourth as compared to conventional tillage. Direct planting without any tillage was practiced as 'no tillage' or 'zero tillage' system. Conservation tillage involves growing of crops with minimum disturbance to soil surface. It was designed to reduce erosion, maintain or improve soil health, and increase infiltration by reducing surface sealing and enhancing macro-pore connectivity. It reduces evaporation by preventing overheating of soil surface, builds up soil organic matter, improves aggregate stability and thus soil structure and helps proliferation of soil micro-flora and fauna and earthworms. Also, it reduced emission of green house gases through use of less amounts of fossil fuel.

5.11.5. Graded farming

Graded farming for high rainfall areas with heavy soil and contour farming for low rainfall areas with light soil are recommended as anti-soil erosion measures. Graded farming meant conducting field operations such as ploughing, cultivating and sowing across the slope on a grade of about 0.2 percent rather than up and down the slope, where as contour farming meant conducting field operations on dead contours. Both these reduced erosion hazard and guided safely excess water through inter-row space. The small ridges formed by intercultural operations and plant stems on a grade guided the runoff water with a non-erosive velocity to drainage channels. Krishnamoorthy and Rao (1974) reported beneficial effects of contour farming to be 40 percent under Bellary conditions and therefore, on black soils with high rainfall, the benefit of graded farming would definitely be 40 percent or more.

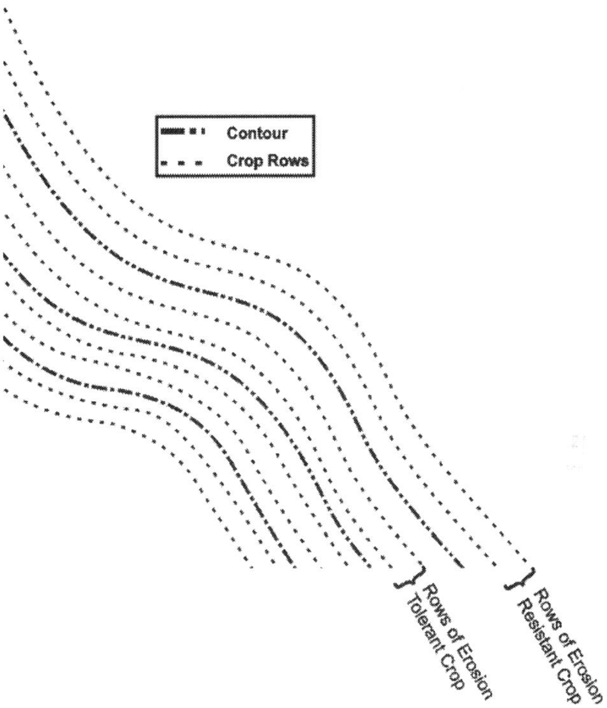

Fig. 16. Strip cropping

5.11.6. Strip cropping

Strip cropping is the practice of growing alternate strips of closely growing and inter-tilled crops in the same field. Under this method, the crop was grown in contour strips at right angle to the slope of the land. The strip intervened or acted as barriers to the flow of water loaded with silt and helped in protecting the soil from erosion. Strips were planted alternately with deep rooted and thick canopy crop, and with thin canopy and fibrous root system crop. For example, legumes and sorghum for fodder and grasses like *Cenchrus ciliaris* (Anjan grass), *Napier grass*, *Pennisetum pedicellatum* (Dinanath), *Dicanthium* sp. (Kale) and *Indigo feraglandulosa* (Bermuda grass) were erosion resistant as they developed maximum foliage at a fast rate. On the other hand, crops with shallow root system and less vegetative growth as compared to legumes and grasses, sorghum (for grain), maize, sesame, pearl millet, etc. were erosion permitting. Therefore for soil conservation the strips of these erosion permitting crops must be alternated with erosion resistant crops (Fig. 16). Strip cropping was more intensive soil conservation practice than graded farming. Strip cropping of green gram, black gram and sorghum was tried in Sagar district during early fifties of twentieth century and became quite popular in Sagar and Hoshangabad districts (Verma 1967).

The strip should be of sufficient width to be convenient for farm operations, yet not so wide as to permit excessive soil losses. The slope of the land and the erodibility of soil should be considered. The approximate maximum width of the strips for different soil and slope conditions was given in Table 24.

Table 24. Strip width for different slopes of a permeable soil

S. No.	Slope (percent)	Strip width (m)
1	2 – 5	30
2	6 – 9	25
3	10 – 14	20
4	15 – 20	15

5.11.7. Alley Farming

Alley farming referred to planting of short duration crops on the land between the rows of long duration plants (trees, bushes, etc.) so as to create appropriate and more effective vegetative cover. This was a tree- cum-crop farming system in which food crops were grown in alleys formed by hedge rows of trees or shrubs. The trees/bushes were selected such that these offered minimum adverse effect on the growth of crops through shading and competition for nutrient-uptake. Also, during growth-period of crops, the trees/bushes were pruned, lopped or cut so as to minimize shading and reduce competition with food crops. The lopping provided green and nutritious fodder for milch animals during lean (dry/scarcity) period. Alley farming when practiced along the contours on a slopy land, provided a barrier to run-off so as to conserve soil and water.

5.11.7.1. Choice of trees: Depending upon the plant species grown, alley farming can be a version of:

i. Agro-forestry (Trees for timber and crops for food)

ii. Silvi-pastoral system (Trees for fuel, fodder and fibre and grasses for fodder, cut-and-carry system)

iii. Agri-horti system (Fruit trees and food crops)

Which of the above systems was suitable at a place determined the choice of tree species. However, the most important criteria for the choice was its amenability to lopping and less competitiveness for nutrients. For example, a leguminous tree due to its nitrogen fixing capacity did not compete for nitrogen and had symbiotic effect. Similarly, hedge forming tree/bush was preferred where the land was slopy and erosion serious.

5.11.7.2. Choice of crops: In an experiment on alley cropping during 1984 to 1986 (three years) at Central Research Institute for Dryland Agriculture, Hyderabad, pearlmillet was found to be most suitable for alleys of *Lucaena leucocephala*, followed by sorghum. As compared to sole cropping, yield reduction of pearlmillet was 30 per cent while that of sorghum 48 percent. The same for pigeon pea and castor was 61 and 62 percent, respectively. However, economic evaluation of sorghum-lucaena and pigeon pea-lucaena systems showed that total income from sorghum - Lucaena alley farming was nearly twice of income from sole cropping of sorghum but total income of pigeonpea–lucaena system was almost seven times the income of sole cropping of pigeonpea (Singh et al 1987).

5.11.8. Mixed Cropping

Here again the purpose was to create appropriate and more effective crop-cover. Mixed cropping of sorghum, maize and pearl millet with green gram, black gram and cowpea etc. provided better crop cover than sole cropping of these and was therefore recommended over sole cropping of millets.

5.12. MANAGEMENT OF FLAT LAND

On flat land (slope less than one percent) water-logging and water-stagnation was a more serious problem than soil erosion, particularly on black soils. The construction of field bunds and property bunds on flat land with clay and clay loam soils particularly led to water stagnation along the bunds and in depressions formed during the construction of bunds. Where topography was flat, problems of waterlogging arose, particularly on deep black (Vertisol) soil having a low infiltration rate. Problems of water stagnation on such soil should also arise if external drainage of the land was restricted. Thus, to solve problems of water stagnation and waterlogging, provision for both the overall land drainage and the crop drainage was necessary, the latter one particularly on deep black soil with flat topography.

5.12.1. Provision of drainage in the inter waterways area

Safe disposal of run-off water from inter waterway area or from fields between two crests to a grassed waterway was also necessary. The construction works were designed according to slope of land e.g. (i) graded bunds on 0.3 percent grade if the slope of land was more than 1 percent but less than 6 percent, and (ii) open channels on a grade (0.2 percent) on comparatively flat lands (slope less than 1 percent). The cross section of graded bunds/open channels would depend upon the rate of runoff and the area of the catchment. Also, depending upon the difference in bed levels of the grassed waterway and the graded bund/open channel, ramps or suitable fall structures had to be constructed.

5.12.2. Management of soil water in root zone

On flat land with black soils and clayey alluvial or other type of soils particularly under high rainfall (greater than 1000 mm), *kharif* crops (except rice) either failed or yielded poorly because of excess moisture in root zone due to saturated soils. The inter channel area (i.e. area between two open drainage channels) on flat lands also needed suitable treatment to manage water in crop root zone, as to rid the active root zone of excess water. For this, one or more of following practices could be adopted depending upon soil conditions and the severity of drainage problem (rainfall, soil depth, slope, land use etc.).

5.12.3. Narrow ridge and furrow system

Preliminary research work at International Crop Research Institute for Semi-Arid Tropics (ICRISAT), Hyderabad, supported the contention that ridge and furrow system could be used to manipulate runoff and reduce erosion (ICRISAT 1974). The ridges laid out along land slope (less than one percent) served somewhat as 'minibunds' and the furrows under a slope served as small but effective surface drains carrying excess water into cut-off drains dug across the slope. This system was very useful on flat lands with heavy soil and under comparatively higher rainfall (1000 to 1200mm). Planting of upland *kharif* crops on ridges, laid out along the general slope on Vertisols of low permeability, increased yields of soybean, sorghum and maize by 30 to 100 percent as shown in Table 25 (Tomar et al 1978, Gupta et al 1979).

Table 25. Effect of ridge planting on crops grown on Vertisol

Crop	Land-slope	Grain yield (kg ha^{-1})		Percent increase
		Ridge Planted	Flat planted	
Soybean	0.3	2900	1890	53
	0.6	2650	1740	52
	1.2	3030	2390	27
Maize	0.3	3376	1601	106
	0.6	3247	1704	88
	1.2	3644	2642	38
Sorghum	0.3	3608	2638	38
	0.6	3754	2445	54
	1.2	3553	2717	30

5.12.4. Broad bed and furrow (BBF) or Bedding System

Research work at ICRISAT recommended BBF system, which involved creation of 90-150 cm wide and 15-20 cm high raised beds with 0.3 to 0.5 percent grade. The beds were separated by 50 cm wide furrows which drained in to grassed water ways. The beds were stable for 2 to 4 years. Observations under All-India Coordinated Research Project on Dryland Agriculture at centres Indore and Rewa showed that 90 to 240 cm broad raised beds (depending on crop and number of rows to be adjusted) made with tool bar attached with furrow openers were easier to maintain and work with and were as satisfactory as narrow ridge and furrow systems. It was worth noting that adoption of BBF system needed a special costly tool bar while broad bed and furrow system could be constructed by just a furrow opener (Fig. 17).

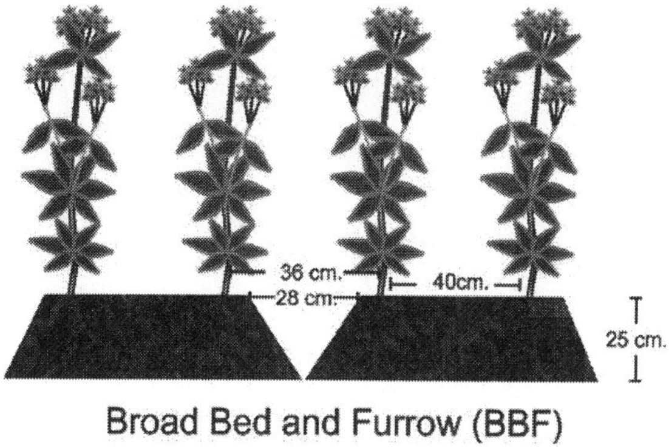

Fig. 17. Diagram of Broad Bed and Furrow system

5.12.5. Graded furrows

Observation in Indo-UK Project area showed that furrow opened on a grade of 0.2 to 0.3 percent served as surface drains and let out excess water safely. The furrow should have cross section of 0.10 to 0.15 m^2 and spaced at 8 to 10 m interval depending upon the crop to be grown and the slope of land (lesser the slope, closer the spacing). Such furrows could be easily constructed by farmers with the help of a desi plough to the share of which a piece of wooden plank had been tied up (Verma, 1982).

5.12.6. Raised and Sunken Bed System

The system developed at JNKVV Jabalpur (Gupta et al 1978) consisted of an array of raised and sunken beds constructed side by side by digging and lifting soil from sunken bed site and depositing it upstream on raised bed site to raise its level and there by create a raised bed. Run-off from raised beds planted to an upland crop like soybean was collected in the adjacent sunken-bed planted to water logging tolerant or standing water loving crop (like rice). Generally the bed-width was kept equal to calculated drain spacing, whereas elevation difference between raised and sunken beds was maintained at 20 to 30 cm. The rainfall and soil characteristics were the deciding factors for the specification of the beds. About 6 to 9 m wide raised beds and six metre wide sunken beds with 25 to 35 cm elevation difference were observed to significantly improve grain yield of upland crops (AICRPDA 1994, Tomar et al 1996). The system was convenient for planting upland crops in rows spaced at 30, 45, 75 and 150 cm. It moderated soil water regime in the root zone and alleviated the adverse effect of poor drainage or soil water stress on crops very effectively. Small earthen bunds on either side of sunken beds ensured uniform spread of run-off in sunken beds. The system facilitated drainage of upland crop (planted on raised bed), provided more water to crop (planted in sunken bed), encouraged in-situ rain water conservation and minimized soil erosion and loss of plant- nutrients (Table 26).

Table 26. Yield of soybean under different land treatments at Indore

Treatments	Yield (kg ha^{-1})	Soil loss (kg ha^{-1})	Nitrogen loss (kg ha^{-1})
Flat (conventional) planting	262	1404	34.9
Broad bed and furrow (BBF)	1333	1332	21.4
Broad bed and tied furrow (BBTF)	1510	717	27.9
Raised and sunken bed (RSB)	1546	221	7.3

Tomar et al (1996) reported that raised and sunken bed system not only improved significantly the productivity of *kharif* crops but also increased *rabi* crop yields. Demonstration of the system on farmers' field recorded more than 100 percent increase in soybean and paddy productivity (Table 27). The system once developed became permanent and needed little maintenance. It had a great potential of replacing 'Haveli system' prevalent on Vertisol in high rainfall areas of M.P. (Gupta, 2002).

Table 27. Effect of Raised-Sunken bed system on soybean and paddy yield

Crop	Grain yield (Tons/ha)		C.D. (0.05P)
	Raised-Sunken bed	Farmers' Practice	
Soybean	3.10	1.15	0.076
Paddy	3.01	1.53	0.062

5.12.7. Tied Ridging

This was modified version of ridge-furrow system, where ridges were tied at 15 to 20 metre interval to allow rain water collection in furrows and its infiltration into root-zone. The system not only conserved rain-water but also substantially reduced the sediment and nutrient losses induced by run-off water. The system was recommended for comparatively low rainfall (less than 700 mm) conditions (Sharma et al 2005).

5.12.8. Broad Bed and Tied Furrow system (BBTF)

This system was similar to Broad bed and furrow (BBF) except that furrows were tied with small cross section earthen bunds at a regular interval of 10 metres, towards the end of rainy season (mid August) so that adequate moisture was available to *kharif* crop during reproductive growth phase. The system not only increased availability of moisture to crop plants, but also reduced the soil loss by sheet erosion to the extent of 40.3 percent as well as loss of nutrients viz. N, P, K and S by 3.5, 13.5, 16.9 and 1.1 percent, respectively. Also, the enhancement in soybean seed yield, water-use by crop and water use efficiency was 17.8, 13.6 and 0.7 percent, respectively due to BBTF system, as compared to conventional system (Sharma 2010).

■■■■

VI

IMPROVEMENT OF PRODUCTIVITY OF RAIN WATER

Till recently the water management was concerned with irrigation water only. With launch of all India Coordinated Research Project on Dryland Agriculture in early seventies, emphasis was laid on rain water management. Pioneer experiments to study rain-water-use efficiency of different crops/varieties and cropping systems were conducted at twenty two research centres situated in different agroclimatic regions of the country. Also, experiments on run-off collection and recycling were started. But real thrust to rain-water management on watershed basis was made when operational research projects on farmers' fields were executed at a number of centres. The first such project entitled 'Indo-British Operational Research Project' (named Indo-UK Dry Farming Project later on) was established by the I.C.A.R. in collaboration with UK Govt. at Indore (J.N.K.V.V. Campus College of Agriculture, Indore, M.P.). The research on run-off management and measurements paid heavy dividends when its results were transferred to farmers' field. Construction of tanks and farm ponds became very popular with farmers, particularly, where water-table had receded considerably due to over-exploitation of ground-water (Verma 2011).

6.1. RAIN-WATER A VERY PRECIOUS GIFT OF NATURE

Rain water is really a boon to all living beings, and the prosperity of a Nation depends on the productivity of rain-water achieved through its appropriate management. Also, if not managed appropriately, it can prove to be a curse, because depending upon the topography and slope of land, and texture, structure and depth of soil, the rain-water creates different types of problems. On flat land with deep black soils having clay to clay loam texture and very low infiltration rate (0.5 to 2.0 mm per hour), rain-water causes problems of water-stagnation and waterlogging leading to even flooding of the area, more so if external drainage of the area is restricted (e.g. in saucer shaped topography). Under such conditions there is excessive amount of run-off with very low velocity of flow. If land-slope is greater than 1 percent, there is excessive run-off with high velocity-greater the land slope, greater the velocity of runoff, causing soil-erosion the severity of which depends on erodibility of the soil. On slopy black soils, shallow (<1m) to deep (>3m) gullies (V-shaped) are formed depending upon the soil depth. On deep alluvial soils, shallow to deep ravines (U-shaped) are formed and the severity of the problem again varies with erodibility of soil. Deep alluvial soil of Chambal valley is very fragile with a high dispersion coefficient and consequently gullies start developing at a fast rate on this soil. Also, subgullies develop soon and a dense network of gullies, subgullies and sub-sub gullies, etc. is ultimately formed as deep ravines and the land surface is visible only as narrow and zig zag high angular bunds. Thus, what to talk of the soil, the entire once cultivated and fertile land was lost to ravines by uncontrolled rain-water run-off. Therefore, the strategy of rain-water management depended upon the nature of problem created by it, which in turn depends on type of land/soil, its topography, permeability and so on. Further, the rain-water has to be managed such that it was available whenever and wherever needed.

6.2. PLENTY OF RAIN –WATER

Fortunately India is blessed with adequate water resources including rain-water. The average annual rainfall varies from less than 20 cms in Western Rajasthan to 120 cms in the plains, 200 cms in West Coast and Western Ghats and as much as 400 cms in southern slopes of Khasi hills and Brahmaputra valley. The country's average annual rainfall is 119.4 cms which is much higher than the global mean of 99.1cms. The total average annual precipitation of the country is estimated to be 394 M ha m. Besides, an annual snow-melt of about 6 M ha m is contributed by permanently snow- clad areas of the country (Verma et al 1996). Thus, the total amount of rain-water comes to 400 M ha m. Of this, 70 M ha m is lost as evaporation and 115 M ha m as surface run off, while the remaining 215 M ha m soaks into the ground as soil profile stored moisture and ground-water recharge. Of this about 45 M ha m regenerates as surface flow (run off in live *nalas*). An additional 20 M ha m is brought in by streams and rivers from catchments lying outside the country in Nepal and Tibet. The Total surface run-off of the country is thus 180 M ha m (Chaturvedi 1985), and the potential water availability, on average, is 57 cms ha^{-1} of geographical area and about 130 cms ha^{-1} of cultivated area. This was an enormous quantity of available water.

6.3. AVAILABILITY OF RAIN-WATER

The picture of rain-water availability was not so rosy as seen from average values of rainfall, because there was tremendous variation in rainfall distribution over time and space. Further considering the mean annual rainfall, only 30 percent area of India received less than 750 mm, 42 percent between 750 and 1250 mm, 20 percent between 1250 and 2000 mm and only 8 per cent received more than 2000 mm rainfall annually. Also, at a particular place, the intensity of rainfall could vary considerably on different days during the rainy season, and there could be prolonged dry periods during the season even in high rainfall areas.

6.3.1 Seasonal variation

The annual rainfall in India had high seasonal variability. The south west monsoon contributed 74 percent of total rain-water. Another three percent was brought by north-east monsoon and the remaining 23 per cent was received through pre-monsoon or post monsoon showers. Thus 80 to 90 per cent of the total annual rainfall was concentrated in four monsoon months. Besides being highly seasonal in nature, the rainfall varied from day to day in a year and also from year to year. There was nothing like normal daily rainfall or annual rainfall. The record of monsoon rainfall was studded with drought and excessively wet years and there was no set pattern of their occurrence.

6.3.2. Variation in number of rainy days

The number of rainy days in the season i.e. distribution of rainfall over time varies considerably from year to year. The distribution of rainfall and not the total amount of rainfall received decides whether a particular year was drought year or not.

6.3.3. Intensity

The intensity of rainfall (mm/hour) was also variable irrespective of total amount of rainfall. Not only in humid regions (receiving >2000 mm rainfall annually) but in semi-arid (rainfall 500 mm to 2000 mm) and arid regions (rainfall<500 mm) also, short spells of high intensity rains occurred.

6.3.4. Duration of monsoon

Duration of monsoon and its onset also varies in time and space. Sometimes monsoon sets in early and continues longer, with an increased duration, while in other years, it may start late, but end sooner and thereby reducing the duration of monsoon considerably.

6.3.5. Total duration of mean annual rainfall

In India, all of the mean annual rainfall is received in about 150 hours of actual rainfall, while half of it descends in not more than 30 hours of heavy spells. Thus, erratic distribution and uncertainty of rainfall add further dimensions to rain-water management.

6.4. STRATEGY OF RAIN-WATER MANAGEMENT

As already stated, the total rain-water received annually in India was 400 M ha m which got disposed as follows:

70 M ha m got evaporated and lost to atmosphere

115 M ha m available as surface run-off

215 M ha m got soaked into ground

Of 215 M ha m of water, 45 M ha m regenerated as run-off, 125 M ha m got stored in soil-profiles and 45 M ha m was underground water recharge. Of these different components, evaporation losses were beyond control, but through appropriate management practices, the surface run-off could be minimised or utilised, and maximum quantity of rain-water could be soaked into soil/land. Thus, the strategy of rain-water management should be: (i) maximisation of in-situ water conservation, (ii) appropriate management of run-off water – a total of 180 M ha m (115 M ha m in India, 20 M ha m from neighbouring countries and 45 M ha m regenerated run-off) on watershed basis, (iii) Increase in use of surface flow through (a) maximum run-off collection into farm ponds, tanks, etc. (b) construction of check dams, submersible check dams on live *nalas* or waterways. (c) through construction of many small dams than a few big dams, and (d) recycling of collected run-off to attain maximum water-use efficiency.

6.5. PRESENT PRODUCTIVITY OF RAINWATER

At present, rain water is handled through following 4 types of farming systems:

1. Open Field Cultivation
2. Haveli System
3. Bandhi System and
4. Bunded Rice Fields.

In all the rain water management systems practiced at present, there was a gross wastage of rain water, particularly under Haveli and Bandhi Systems, in which vast areas were still left fallow in *kharif* season when more than 80 per cent of total annual rainfall was received. Even under open field cultivation system, part of the land was left fallow resulting in not only wastage of rain-water but also of soil through water erosion. *Kharif* fallow under open field conditions increased as a result of construction of field bunds on black soil, in the name of contour bunds by soil conservation service (Verma 1994). The extent of *kharif* fallow in M.P. was shown in Table 28.

Under open field cultivation system, either one year rotation: *kharif* crop-fallow or two year rotation: fallow *rabi* crop – *kharif* crop- fallow was practiced, irrespective of land slope, soil depth and texture. Thus, only one poor crop a year, either in *kharif* or in *rabi* was grown. Further, the soil was continuously impoverished by water erosion, particularly under *kharif* fallow, and lack of fertilizer application to *kharif* crop if grown. Apparently, the productivity of rain water was the lowest over large areas where *kharif* fallow was practiced. In *haveli* system, there was no soil erosion due to flat topography and bunded conditions, but productivity of rain-water was limited to that small fraction of it which got stored in the soil profile and used by crop plants for their growth during *rabi* season. That is, the productivity was very low. In bandhi system, no use of enormous quantity of rain-water impounded against bandhis was made, not even for life saving irrigation to *kharif* crops grown on upper part of the slope. On submerged land *rabi* crop was grown after draining away impounded water and thus only profile stored moisture supported the *rabi* crop, resulting in very low productivity of rain-water again. In low lying areas and high rainfall regions, field bunding has been done to create rice fields. Also, these bunds have become fully stable by now and hold water even if excess run-off spills over these bunds. While the productivity of rain-water on growing *kharif* rice crop was more than other systems, construction of rice-fields on very light and shallow soil was not remunerative as in spite of high rainfall the yield of rice crop was poor due to dry periods between the rains. As a matter of fact, such areas are drought prone only due to rice crop being grown there. Thus, on the whole, the productivity of rain-water was very low at present.

Table 28. Extensive Fallow in *Kharif* Season (1978-79)

S. No.	District	Area Cropped in *Kharif*		Area left fallow (% of total cultivated)
		000 ha	% of cultivated	
1	Raisen	81.9	19.9	80.1
2	Vidisha	118.0	22.6	77.4
3	Hoshangabad	109.8	23.9	76.1
4	Sagar	135.1	25.4	74.6
5	Bhind	101.9	29.5	70.5
6	Guna	152.0	32.2	67.8
7	Gwalior	91.7	33.4	66.6
8	Datia	46.3	34.8	65.2
9	Morena	145.5	35.0	65.0
10	Narsinghpur	101.4	35.6	64.4
11	Damoh	107.6	36.4	63.6
12	Satna	154.3	37.6	62.4
13	Bhopal	50.2	39.1	60.9
14	Panna	92.8	39.3	60.7
15	Rewa	214.9	46.8	53.2
16	Chhatarpur	180.5	47.8	52.2
17	Sehore	179.8	47.9	52.1
18	Jabalpur	222.6	48.4	51.6

Source: Agriculture Statistics 1980. Directorate of Agriculture, M. P. Government, Bhopal (M.P.)

6.6. INCREASING PRODUCTIVITY OF RAIN WATER

In all the rain-water management systems discussed above, a gross wastage of rain-water takes place particularly under *haveli* and *bandhi* systems in which vast areas are still left fallow in *kharif* season when more than 80 percent of total annual rainfall was received. Therefore, the first step to increase the productivity of rain-water was to grow crops over as large areas as possible and thereby reduce *kharif* fallow to the extent possible. Also, a well nourished *kharif* crop with good canopy provided a very good vegetative cover to protect the soil against water-erosion. Further, soybean crop provided a better soil-cover than maize, sorghum, pearlmillet and so on. Thus, on land/soil with slope >one percent, growing a *kharif* crop not only increased the productivity of rain-water but also protected the soil against water erosion.

6.6.1. Introduction of soybean

The experiments conducted under All India Coordinated Research Project, Indore, (1971-73) showed that soybean was the most remunerative crop under rainfed conditions on black soil. The same was demonstrated on farmers' fields under Indo-UK Operational Research Project (1975-1979) starting with replacement of *Kalitur* (Black soybean) by T-49 variety of soybean and introducing other high yielding varieties of soybean later on. Gradually, soybean became very popular particularly on black soil and spread like wild fire making M. P. the soybean state.

6.6.2. Reduction in *kharif* fallow

Along with soybean, hybrid varieties of sorghum and high yielding varieties of other *kharif* crops e.g. pearlmillet, blackgram, greengram, pigeonpea, sesame etc. were also introduced and farmers soon realized that under rainfed conditions it was more paying to raise crops in *kharif* season against their earlier opinion that growing crops in rabi was more dependable and profitable. Consequently, *kharif* fallow in M. P. got reduced with time as shown in Table 29.

Table 29. Reduction in Area of *Kharif* Fallow

Year	*Kharif* Fallow (lakh ha)
1978-79	51.81
1979-80	45.40
1980-81	46.42
1981-82	35.33
1982-83	37.97

6.6.3. Increasing productivity of rain-water through improved package of practices

In *kharif* season, moisture is usually not a limiting factor, and therefore adoption of improved package of practices like selection of appropriate crop and its varieties, balanced fertilization based on soil test through integrated nutrient management, interculture, integrated pest management and harvesting crops at physiological maturity stage can be profitably practiced.

6.6.4. Increasing productivity of rain-water on flat land with deep soil

On flat land, the movement of rain-water (run-off) was very slow, particularly on deep black soil which had very low rate of infiltration. Consequently, the problem of water stagnation/waterlogging was serious, more so under bunded conditions. The problem was not serious on flat land with sandy loam and loam soils, even under bunded conditions because such soils were well drained having high infiltration rate. Water stagnation/waterlogging on these soils was temporary and most *kharif* crops could withstand it. Haveli system is presently followed on deep black soil receiving high annual rainfall. JNKVV scientists have developed 'Raised and Sunken Beds Technology' to raise *kharif* crops viz. soybean and rice on raised (3 m wide and 20 cm high) and sunken beds (6 m wide), respectively. Paddy strips served the purpose of interplot water-harvesting. The system is found to be most remunerative and most efficient in increasing the productivity of rain-water. However, it needs to be demonstrated on farmers' field for adoption.

6.6.5. Increasing productivity under Bandhi system

Under Bandhi system, run-off was collected against bunds but it was not recycled. A system of independent surplussing arrangement of run-off water by constructing a grassed waterway perpendicular to bandhis and its collection in a suitable way at appropriate point should enable its re-use when needed (Verma 1990).

6.6.6. Double cropping

Another advantage of raising a crop in *kharif* instead of leaving the land fallow was that double cropping was possible under raifed conditions, particularly if a short duration crop/variety was grown in *kharif*. Usually there was adequate moisture in lower layers of a deep soil, particularly under black soils, after the harvest of *kharif* crop. The chances of double cropping increased if seed could be placed in moist zone without tillage (zero tillage), and if there was a late rain to moisten the surface soil. Sharma and Thukral (1987) studied the effect of rainfall distribution on the possibility of double cropping under rainfed conditions and observed that due to early withdrawal of monsoon in 1980-81 and 1984-85, seeds of second crop of chickpea and safflower (after soybean and maize in *kharif*) could not germinate as the moisture content of surface soil (0-15 cm) was not sufficient. But in 1981-82 and 1983-84, some rainfall after the harvest of *kharif* crop enabled the seeds of second crop to emerge. In the year 1982-83, the total rainfall received during July-September was only 533.9mm which was less as compared to 1980-81 and 1984-85, but 41.2mm and 56.6mm of rains received in second fortnight of October and first fortnight of November, respectively, resulted in successful double cropping under rainfed conditions.

6.7. RUN-OFF COLLECTION AND RECYCLING

Run-off collection and recycling was the most important step to be taken for increasing the productivity of rain-water. Also, one of the most important elements in the design of improved soil and water management techniques for the semi-arid tropics was the concept of harvesting excess runoff from small agricultural watersheds in water storage reservoirs for subsequent use in supplementary irrigation of upland crops. In many areas of India, a rather low percentage of annual rainfall was available and used actually for crop production due to lack of appropriate management of rain-water. Consequently, water became the most important limiting factor for the productivity of crops and made the rainfed agriculture risky. Thus, the first and foremost consideration for the improvement of rainfed farming was to manage rain water so as to conserve and efficiently utilize the rain water. Every effort should be made to conserve as much rain water as possible, in situ i.e. right where it fell by manipulating the surface soil, using surface mulches, stubble mulches and through tillage to modify land configurations. Inspite of all these efforts, runoff was still a certainty and the emphasis should be given on runoff collection and its reuse.

6.7.1. Water harvesting

Water harvesting was a practice which encompassed all those techniques, methods and innovations which induced collection and storage of water from various sources for its beneficial use. It was actually collection of run-off water from small natural catchments and recycling it to meet the demand of water for various purposes including agriculture (growing crops, trees or for livestock consumption). Sometimes a distinction was made between 'water harvesting' and run-off collection on the basis whether a natural catchment or an artificially treated catchment was used for collecting the run-off. Water harvesting sometimes implied the run-off collection from artificially treated or specially prepared catchments. The treatment might be (i) soil compaction, (ii) making the interplot space slopy, or (iii) any other treatment such as spreading canvass, plastic sheet or metal sheet resulting in greater amount of run-off. However, water harvesting must be practiced with an integrated management of catchment so as to make the best possible use of land and water resources available in a catchment.

6.7.2. Characteristics of water harvesting methods

Various characteristics of water harvesting/run-off collection were-

i. Water harvesting and run-off collection were normally practiced in arid and semi-arid regions, respectively, where run-off was of an intermittent character. The surface water in the area flowed for a part of the year and stopped during dry period. Because of limited flow spells, storage in some form became inevitable.
ii. The practices of water harvesting depended on local water resource such as surface run-off, creek flows and springs and therefore did not include storing river water in large reservoirs or mining of natural ground-water aquifers.
iii. These were normally limited or small scale operations in terms of catchment area, volume of storage and capital investments.

6.7.3. Area of applicability

The practice of runoff collection and recycling could be followed in areas where the annual rainfall was more than 500 mm, rainfall intensities were higher than the infiltration rate and the rains were prolonged. Thus semi-arid, sub-humid and humid areas were quite suitable for run-off collection and recycling. The success or failure of rain-water harvesting in arid and semi-arid areas depended to a great extent on the quantity of water that could be harvested under given conditions of rainfall and catchment characteristics. The relevant catchment factors were: surface storage, rainfall intensity and infiltration capacity of soils of the catchment. The run-off efficiency of a catchment was the ratio of run-off volume and precipitation volume. The run-off efficiency was expressed as annual average. It depended on conditions such as extent of natural surfaces, slope of natural surface, extent of vegetative cover (bare or partly or wholly covered), antecedent soil moisture, rainfall intensity, storm duration, catchment size, treatment given, number of years passed after the treatment, etc. The reported run-off efficiencies of natural catchments varied from 20 to 80 percent, while efficiency up to 96 percent was reported for treated catchments.

6.8. RAINFALL AND RUN-OFF

The quantity of run-off at a particular place varied according to land characteristics, soil properties, terrain conditions and prevailing rainfall characteristics (total rainfall, rainfall distribution and rainfall intensity). While there were opportunities for run-off collection under varying situations, terrain conditions and rainfall characteristics in black soil region were highly conducive to generation of high volumes of run-off. In different parts of M.P. and Chhattisgarh states, an annual rainfall of 650 to 1600 mm was received. Even in drought years, the intensity of a few showers was such that at least a part of rainfall was definitely available as run-off, particularly on black soils. Rainfall in Malwa region of M.P. normally varied from 700 to 1200 mm with an average of 900 mm received during mid-June to September end and characterized by a few intense storms. Three rainfall peaks were common in normal years: one in the 28th standard meteorological week (9th -15th July), 2nd in 31st meteorological week (30th July – 5th August) and 3rd peak in 35th meteorological week (27th August to 2nd September). During these weeks storms were not only intense but continuous also. On a 1:10 years frequency, the maximum intensity of rainfall varied from 167 to 245 mm hr^{-1} for 5 minutes and from 67 to 100 mm hr^{-1} for 60 minutes frequencies at Indore. The prolonged rains with high intensity caused large amounts of runoff. Further, by end of July, surface soil in areas with Vertisols and associated soils got completely saturated and their rate of infiltration decreased considerably and therefore, runoff volume increased tremendously.

Run-off measurement studies on small agricultural watersheds under All-India Coordinated Dry land Research Project during 1973-87 (Shrivastva and Verma 1975, Shrivastava 1987, Verma and Gupta 1988) indicated that 21 to 53 percent of the total rainfall was available as run-off on medium black soils and less than 1 to 24 percent on deep black soils at Indore (Table 30 and 31). At Rewa, at least 30 to 40 percent of total rainfall was recorded as run-off (AICRPDA 1980). At Jabalpur, the run-off observed was 64 percent from maize and 54 percent from soybean field (Bhargava et al 1973). Sharma (2010) studied run-off from different land treatments (flat, broad bed and furrow, broad bed and tied furrow and raised-sunken bed) on deep black soil under soybean crop at Indore during 1988-97, and reported minimum run-off under raised-sunken bed system. The effect of other land–treatments on run-off was not found to be significant (Table 32).

Table 30. Rainfall and Runoff from medium black soil at Indore

Year	Rainfall (mm)	Runoff (mm)	Runoff (percent)
1971	975	523	53.6
1972	539	184	34.2
1973	1889	936	49.0
1974	868	182	21.0
1975	1376	398	28.8
1976	1241	524	42.2
1977	1131	417	36.9
1978	785	389	39.5
1979	675	270	40.0

Note: Area of runoff plot = 0.1 ha, Slope of the plot = 0.6 percent

Table 31. Rainfall and Runoff from deep black soil at Indore

Year	Rainfall (mm)	Runoff (mm)	Runoff (percent)
1979	675	140	20.72
1980	973	165	16.93
1981	840	94	11.20
1982	629	4	0.7
1983	987	40	4.0
1984	879	206	24.3
1985	636	17	2.7
1986	928	201	21.7
1987	762	48	6.3

Table 32. Run-off from different land treatments (mean of 1988-1997)

Treatment	Rainfall (mm)	Run-off (mm)	Percent
Flat (untreated)	831	111	13
Broad Bed and Furrow	831	96	12
Broad Bed and tied Furrow	831	116	14
Raised-Sunken Bed	831	66	8

Studies conducted recently on farmers' fields under Operational Research Project (Gauli Palasia micro-watershed, Indore) revealed that the run-off from treated section of the watershed was less than that from untreated section and that the land treatment undertaken by the Project increased the underground water recharge (Table 33)

Table 33. Run-off from Gauli Palasia (Indore) micro-watershed (Sharma, 2010)

Year	Rainfall in recorded events	Run-off from treated section		Run-off from untreated section	
		mm	percent	mm	percent
2003	120 mm	20.7	17.3	42.7	35.6
2004	362 mm	35.0	9.7	134.0	37.0

Thus a large quantity of rainwater was available for collection and using it for charging the underground water as percolation tank or for recycling it for irrigation by individual's farm ponds or large community tanks depending upon the site. This could also be achieved by putting stop dams and submersible check dams across the natural streams wherever the situation permitted. Rejuvenation of already existing tanks (now out of order) was also necessary. All these works increased the rate and quantity of infiltration of water into the soil and ultimately increased the recharge rate of existing irrigation wells in the surrounding areas. The runoff water collected in tanks/farm ponds could be used not only to save *kharif* crops in drought years but could also provide a come up irrigation for planting a second crop in *rabi* after the harvest of *kharif* crop.

6.9. PRACTICES OF WATER HARVESTING/RUN-OFF COLLECTION

Practices of water-harvesting were specialized and location-specific for efficient and cost-effective run-off collection. It had to be on the basis of distinct watersheds or subwatersheds according to land capability, topography, environmental conservation and relevant local factors. These included:

i. Run-off diversion for storage or directly irrigating crops,
ii. Construction of check dams and gully plugs of appropriate designs and materials,
iii. Construction of farm-ponds/tanks,
iv. Water spreading techniques for recharging ground water, construction of percolation tanks or recharge tanks,
v. Construction of trenches, pits or small storage along contours in waste-lands, range lands or on bare hill slopes, so as to retain run-off.
vi. Interplot water harvesting where one plot/strip was treated suitably (compacted, treated with sodium salt/paraffin wax or asphalt) so as to induce run-off to the adjoining plot/strip. This was practiced in arid regions only and was not required in Central India.

6.9.1. Construction of Percolation Tanks/Recharge Tanks

These are being used commonly for water harvesting on favourable sites. These, by definition, are small tanks or reservoirs intended for harvesting the run-off in monsoon or wet season, and using the storage exclusively for recharging the substrata. The most important consideration was that the farmers below should be capable of using the recharge done by the water stored temporarily. Percolation tanks constructed over very imperious bed or sheet rock did not serve the purpose of re-charging. Over designed tanks may not need surplussing arrangement but depending upon local conditions, these may be needed. Silt brought in from the upper catchment would render the bed impervious in the long run and this must be removed when needed and used for fertilization of adjoining agricultural lands. Alternatively, recharge tanks might be converted into regular tanks to store run-off. Percolation tanks can also be planned as temporary storage tanks for facilitating irrigation to *kharif* crops by lifting water suitably.

6.9.2. Trenches, pits or small storage along contours

Contour trenching, contour bunding, pits or small rectangular/trapezoidal storages, dug along contours can be used for harvesting run-off and transferring the same to underground strata. Later on recharged water can be used by lifting it. A new measure known as "conservation ditching" (trapezoidal ditches laid on contours) has been tried at Bellary, on deep black soils, successfully.

6.9.3. Submersible check dams

Submersible check dams were constructed in *gullies/nalas* to stabilise these by preventing their bed-corrosion through deposition of silt on the bed. Depending upon the nature of the *nala*, water stood against the check dams for varying periods. If the bed of the *nala* was pervious, the water percolated to charge the ground water. With time, silt clogged the pores of permeable bed and thus water collected against the check-dam (Figure 18). Depending upon the need, the water could be used for irrigating *kharif* crop or as pre-sowing irrigation for *rabi* crop.

6.10. SITE OF TANKS AND FARM PONDS

In Madhya Pradesh and Chhattisgarh states, large community tanks (small reservoirs) had been used to collect run-off and recycle it for centuries. However, privately owned tanks were popular only where underground water was rather limited. Also, these were more promising on deep black soils with high rainfall. Extensive studies on run-off collection through construction of tanks were initiated under All India Coordinated Research Project, main Centre Indore, as early as 1973-74 at College Farm and on farmers' fields (Figure 19) later on (1975-81). Considerable expertise had been gained with regard to selection of site, soil conditions, shape of tanks and design of tanks.

6.10.1. Criteria for site

Various criteria for selecting the site of a farm pond were: (i) its catchment size, (ii) minimum earth work, and (iii) ability to irrigate fields below by gravitational flow and upland fields by lifting water through pumps. If earth work (digging) was involved, the economics (cost of water per unit amount) depended upon excavation and storage created which was site specific. This necessitated the pond to be so located that it had maximum run-off flowing into it. From economic point of view, a pond should be located where the largest storage volume could be obtained with least amount of earth work. Earth work (Ew), storage (S) ratio should be as minimum as possible. Ew/S=1 had the lowest efficiency. Construction of tank at the lowest end of the water way can reduce Ew/S ratio by providing above ground storage. The head of a gully was preferred as the site of farm pond so that it prevented retrogression of the gully, and provided storage of water at higher elevation comparatively. A site at the tail end of a sub-catchment was very suitable for a pond as the 'run-off' from the whole sub-catchment would be usefully collected. The lowest lying field of a farmer's holding at his farm-boundary also made a good site for a farm pond where it could be easily and economically (with suitable excavation: storage ratio) constructed by raising one bund of the field or at the most three bunds of the field, using the soil from excavation.

Fig. 18. Submersible Checkdam in Rainy Season

6.10.2. Vicinity of village

Provided the site was otherwise suitable, it should be located near the village so that it served as the source of drinking water supply to the village cattle. Although the dimensions of the pond would depend upon the site, a deeper pond with a small exposed surface had much less evaporation losses and consequently, longer availability of the water supply to the cattle. On such ponds, a paved ramp on a suitable side was made for the cattle to reach water. Also, a suitable spill-way was necessary, so that excess of run-off water passed away safely when the pond was full. In areas of shallow black soil, a pond/tank served only as percolation tank for a long period of about 10 years, as tank bed had weathered basalt rock-permeable *murrum*. However, at the end of ten years, the percolation rate was reduced to 2mm from original rate of 10mm hr^{-1}. A major handicap to the long life of these farm ponds was their silting up at a fast rate. To overcome this problem, the inlet to the pond should have a gabion structure to trap the silt upstream and allow only water to enter the pond.

Fig. 19. Run-off Collection Tank

6.10.3. Soil Condition

At Agriculture College farm, Indore, two tanks dug on shallow black soil served only as percolation tanks for a long period of about five years, as tank bed comprised of weathered basalt-permeable *murrum*. However, at the end of five years, percolation rate was reduced to 2mm hr^{-1} from the original rate of more than 10mm hr^{-1}. This was due to blocking of pores of *murrum* by silt particles brought by run-off water and deposited into tank every year. Another tank on slopy (three percent slope) medium black soil constructed by excavating lower part of the field and putting the excavated soil along the adjacent bunds to make a strong bund on lower side and tapering side bunds perpendicular to it on both sides proved very successful. It stored run-off water which was used as pre-sowing irrigation for *rabi* crop leading to raising of double crops under rainfed conditions. Thus, type of tank constructed depended on soil type, and medium and deep black soils (Vertisol) were better suited for construction of tanks for run-off collection and recycling.

6.10.4. Shape of tanks

Tanks of different shapes and sizes were constructed on different farmers' fields under Operational Research Project on Dryland Agriculture, Indore. Depending upon the site, rectangular, square and circular shaped farm ponds/tanks were dug on different holdings. Rectangular shape was found to be most popular as these could be constructed easily by excavating along the lowest field-bund and putting the excavated soil on the lower bund and on both the bunds perpendicular to it. Thus, capacity of the tank was much higher than the excavated part. The bunds must be stabilised by compacting the soil and planting grass or other vegetation on the bunds. The run-off flowing in the form of sheet into tank was bound to erode the excavated bank of the tank due to march of gully heads against the flow of run-off. If this was allowed to continue, the bank will disappear leaving a slopy run-off sheet and lot of soil will get deposited into the excavated farm pond/tank. Therefore, appropriate management of the run-off running into tank at several places, was necessary by guiding it against a stable bund (constructed above the excavated bank) and allow it to enter through a proper outlet in the bund. Rectangular tanks are easy to construct in a lowermost slopy field also by excavating as above and putting the excavated soil on the lower field boundary to make a strong bund and on two side boundaries to make tapering bunds. Apparently, the excavation and consequently the storage capacity of the tank will be limited and submergence area will be comparatively higher. The circular tank though needing minimum area for a particular size, was a bit difficult to dig and was not liked by farmers due to difficulties faced by them in using the patches of land around a circular tank for raising crops.

6.11. DESIGN OF TANK

Storage capacity of tank was decided on the basis of available run-off, area to be irrigated and expected losses of water. It was not economical to design tanks for storing total run-off from a catchment because the intermittent nature of run-off increased the effective storage period and seepage losses. Capacity should be designed in such a way that tank was full at the end of monsoon. An ideal basis of designing a tank was to analyze the distribution of rainfall occurrence and seepage and evaporation expected from the tank so that the tank gets completely filled with the last monsoon shower. It was always safe to design the tank on the basis of runoff data from relatively drier years and provide suitable spillway. In the absence of any impermeable soil layer, depth of tank should be kept 2 to 5 metres. Side slope should be a little flatter than the angle of repose for stability of banks of a tank. The containment of the tanks could be increased by increasing depth. At Rewa, a pond/tank constructed by excavation in the lower corner of a field was very successful on flat land. On slopy land, run-off collection was very common by constructing bandh/bandhis against slope.

6.12. SPECIFICATIONS FOR MALWA REGION

The specifications for tank in Malwa Region of M. P. are suggested as follows:

i. Depth of tank : 2 to 5 m

ii. Side slope : 1:1

iii. Tank capacity

   (a) For medium black soil-1500 cubic metres per hectare of the catchment

   (b) For deep black soil–1000 cubic metres per hectare of the catchment

(iv) The ratio of the tank area to the catchment area should be 1:10 or 1:16. The area under tank can be reduced by increasing the depth of tank.

(v) Grass filter strip: The grass filter strip, about 30 m long should be provided in the waterway just above the pond to reduce the velocity of run-off water if the tank was situated at th end of waterway. This will also allow the silt to settle on the bed of waterway before entering the pond.

6.13. SILTATION OF TANKS/FARM PONDS

As already stated, fast siltation of farm ponds/tanks was a serious problem and more so in fragile Entisols/Inceptiols where water stable aggregates percentage was low and dispersion coefficient high, resulting in high silt load of run-off water. Another important factor affecting the silt load was the condition of the catchment area of the tank. Maximum silt load was observed when slopy ploughed land was left fallow in *kharif*, i.e. when fallow-*rabi* crop rotation was followed by the farmers under rainfed conditions. An interesting study was conducted under Operational Research Project, Indore, during 2005-08 to see the effect of (i) type of machine used for construction of the tank, and (ii) Shape of tank (rectangular vs circular) on extent of silting during monsoon season. It was concluded that excavated tanks constructed by using JCB machine had greater silt deposit than those constructed by using bulldozer, because satisfactory compaction of tank bunds could be achieved only by using bulldozer and not by JCB. Also, where excavated soil was spread on the fields around the tank, it resulted in greater silt-deposit in the tank, for obvious reason. Further, the channelized flow into the tank through earthen channel caused greater silt deposit in the tank (Jain *et al* 2014). This will happen only if the earthen channel was not stable and both the sides and the bed of the channel were eroded by run-off water. The walls (banks) and bed of the channel must be made stable by vegetation or suitable mechanical structure where run-off water entered the tank. Similarly, run-off water coming from above in the form of sheet-flow could damage the particular tank-wall if it kept on falling from top of wall into the tank. Actually, this could result in collapse of the wall just like advancement of gully head unless steps were taken to prevent it. Similarly, it was not understood how and why circular tanks got less silt-deposit than rectangular tanks if appropriate measures were taken to stabilise the inlet in both cases and also proper side-slopes were maintained in both types of tanks. It was easier to maintain appropriate side-slope in case of rectangular tank than circular one. Irrespective of shape of tank if the sides (walls) of tank are vertical, these are likely to fall off to attain a natural angle of repose. Thus, sides (walls/wall) of tank should be flatter than natural angle of repose, to be stable.

6.14. CONSERVATION OF STORED WATER

Stored run-off water was to be used for both crop production and livestock. It was necessary to conserve it so that maximum water was available when needed. Conservation of stored water was accomplished by minimizing the losses through evaporation and deep percolation.

6.14.1. Losses due to evaporation

Significant amount of water was lost from water bodies through evaporation. Measures to reduce evaporation losses include:

i. Formation of mono-layers over water surfaces by chemicals e.g. cetyl/stearyl alcohols,
ii. Using stored water as early as possible so as to cut down the time of evaporation.

6.14.2. Losses due to percolation and seepage

The losses due to seepage and percolation were estimated to be 0.4 mm per m^2 of the wetted area of the tank per day on deep black soil and 2.0 cm per m^2 for permeable *murrum* bottom. Many sealing materials for seepage reduction from water harvesting tanks have been suggested. These included bentonite, concrete, plastic film (silpaulin), brick lining, low density polyethylene film (L.D.P.) overlaid by brick lining, sodium salts and locally available clay and cow dung slurries. The selection of a particular lining material for any area required complete analysis of tolerable seepage, cost of maintenance, availability and effective life of lining material. At Indore, the reduction in seepage losses for constant head of 1.45 m was almost 100 percent when plastic film (silpaulin) was used as lining material. L.D.P. film overlaid by brick lining was found to be next best and losses were reduced from 20 to 6 litres (86 percent reduction) per square metre of the submerged area per day, followed by cow dung-straw-soil mixture which reduced seepage losses by 82 percent. Of other sealants, soil-cement mixture reduced losses by 44 percent while bentonite was ineffective.

6.15. RECYCLE OF COLLECTED RUN-OFF

Surface stored water can be used for not only life saving irrigation to *kharif* crop but also for pre-sowing irrigation to *rabi* crop to achieve double cropping under rainfed conditions. This has been proved by results of experiments conducted under All India Coordinated Research Project on Dryland Agriculture Centre Indore, at the research station and also demonstrated on farmers' fields under Operational Research Project, Indore.

6.15.1. Effect of life saving irrigation

The effect of life saving irrigation was studied at Indore in 1978 when monsoon rain ceased on August 31. The yields of soybean (variety Bragg), maize (Variety

Ganga-5) and black soybean (variety JBS-1) increased by 14.2, 17.0 and 7.0 percent, respectively. The short duration varieties of soybean (JS-2) and sorghum (CSH-5) did not require the irrigation (Table 34) and produced normal yield escaping the drought but long duration varieties of soybean and maize gave increased yields only when provided with one irrigation. Also, sorghum variety Vidisha 60-1 produced only fodder and no grain, without irrigation, as it was long duration (160 days) variety.

Table 34. Effect of life saving Irrigation to kharif crops

Crop/Variety	Average yield (q ha^{-1})		Increase in yield (percent)
	Without irrigation	One 8cm-irrigation	
Soybean JS-2	15.8	Not needed	-
Soybean Bragg	17.9	20.5	14.2
Maize Ganga-5	29.5	34.5	17.0
Sorghum CSH-5	29.3	Not needed	-
Black soybean JBS-1	14.8	15.8	7.0

6.15.2. Double cropping under rainfed conditions

As a result of the life saving irrigation applied to *kharif* crop, the moisture content of the soil after harvest of the crop was adequate enough to raise a second crop of safflower and gram in the same field. This advantage of double cropping could also be availed under rainfed conditions by providing just one irrigation through recycling of run-off (Table 35). Where such irrigation was not available, what to talk of double cropping, even full yield of kharif crop alone was not obtained when monsoon rain ceased too early. However, full yield of *kharif* crop could be taken by adopting short duration varieties of soybean and sorghum. Thus, it was advisable to grow short duration crops/varieties in *kharif* under rainfed conditions so that full yield of at least *kharif* crop was assured and double cropping could be practiced if there was a late rain or one irrigation was available from run-off collected rain-water.

Table 35. Yield of double cropping after life saving irrigation to kharif crops

S.No.	Crops		Grain yield (q ha^{-1})	
	Kharif	*Rabi*	*Kharif*	*Rabi*
1.	Soybean	(i) Safflower	15.54	15.31
	(JS-2)	(ii) Chickpea	15.54	8.49
2.	Soybean	(i) Safflower	20.50	12.90
	(Bragg)	(ii) Chickpea	20.50	5.82
3.	Maize	(i) Safflower	34.50	18.59
	(Ganga-5)	(ii) Chickpea	34.50	9.79

6.15.3. Effect of supplementary irrigation

Normally monsoon rain-water is sufficient to meet the requirement of *kharif* crop. Therefore, run-off stored water could be utilized for one or two supplementary irrigations to *rabi* crops. A study on the effect of supplementary irrigation to various crops was undertaken at different centres of AICRPDA and results presented in table 36 (Singh 1983).

Table 36. Effect of supplementary irrigation on yield of various crops

Center	Crop	Yield (q ha^{-1})		Percent increase
		Without irrigation	With one irrigation	
Agra	Wheat	21.9	27.4	25.1
Anand	Bidi tobacco	12.1	18.1	49.6
Bangalore	Finger millet	16.2	23.2	43.2
Bellary	Sorghum	9.3	13.7	47.3
Bijapur	Sorghum	16.5	23.6	43.0
Dehradun	Wheat	21.4	35.5	65.9
Hyderabad	Sorghum	13.9	19.2	40.1
Indore	Maize	29.5	34.5	17.0
	Soybean	17.9	20.5	14.5
	Wheat	12.5	17.0	36.0
	Safflower	12.0	15.3	27.6
Ludhiana	Wheat	19.2	41.1	114.1
Ranchi	Barley	5.5	7.0	27.3
Rewa	Upland rice	16.2	27.8	71.6
	Wheat	9.7	18.8	93.8
Varanasi	Barley	26.0	33.6	29.2

It is evident from Table 36 that just one irrigation (5 to 8 cm) to *kharif* (when needed) and *rabi* crops brought out significant yield increases. The greatest response to supplementary irrigation as well as the largest variation in the response was noted in case of wheat-only 25.1 percent increase at Agra while 114.1 percent at Ludhiana. The increase in yield of sorghum was 40.1 percent at Hyderabad, 43.0 percent at Bijapur and 47.3 percent at Bellary. The lowest response (14.5 percent) was by soybean at Indore, followed by maize (17 percent) also at Indore. Good response (71.6 percent) by upland rice crop was obtained at Rewa.

6.16. RUN-OFF RECYCLING AND PRODUCTIVITY OF RICE

In Chhattisgarh state rice was grown on level land with small bunded fields while in M.P. it was grown on different types of soils with varying depth and slopes and receiving different amounts of rainfall. Some regions in M.P. are chronically drought affected only because rice crop grown by farmers was not appropriate for his soil conditions. In spite of high annual rainfall, adequate quantity of water was not always available to rice crop due to erratic distribution and long dry spells between two rainfall events. Consequently, the crop either failed completely or produced very poor yields. In Dindori and Mandla districts of M.P. both the topography of the land and the total annual rainfall were appropriate for run-off collection. Under '*Tar* bund and *Dabari* scheme' a large number of farm ponds and tanks were constructed on farmers' fields and as a result of this irrigation facility, not only rice yields were increased considerably but cropping intensity also was higher, as shown by the data presented in table 37 (NATP 2001).

Table 37. Effect of run-off recycling on cropping intensity and yield of rice (District Dindori, M.P.)

Farmer	Capacity of tank (m^3)	Yields of rice (q ha^{-1})			Cropping intensity	
		Without irrigation	One irrigation	Increase (percent)	Without irrigation	Supplementary irrigation
Sat Singh	2550	552	1304	136	100	200
Jham Singh	1603	623	1219	96	100	200
Hira Singh	1656	887	2098	137	100	200
Ram Singh	2744	996	2007	102	100	200

6.17. FARM POND/TANK BASED INTENSIVE FARMING SYSTEMS

Besides increasing the productivity of various crops and cropping systems, the run-off rain water collected into farm ponds/small tanks could enable the farmers to adopt various intensive farming systems recently discovered. The recent integration of fish with the livestock and crop not only helped to improve the fertilizer and feed supplies, but also increased the farmers' income considerably due to high market value of fish as feed and/or food. Studies conducted in different parts of the country have revealed that such integrated farming systems not only increased the income of farmers many folds but also envisaged harnessing the synergies among different agricultural subsystems/ enterprises augmenting the total productivity, profitability, sustainability and gainful employment of farmers in general and of small and marginal farmers in particular (Singh and Gangwar 2011). Alternatively, the rice-field-pond system can be used to grow aquatic crops along with rice. These different mixed farming systems can be easily adopted in middle Indo-Gangetic basin (Eastern U.P. and Bihar), Chhattisgarh and M.P. (rice tract).

6.17.1. Rice-Fish-Azolla farming system

In this system, rice crop and fish were raised together simultaneously in rice-field/fields having a farm pond/tank in the lowest field. The pond/tank size was approximately one tenth of the rice area and its depth could be adjusted according to the quantity of run-off expected. The tank/ pond was constructed by piling up the dug soil on the lowest side to a desirable height and on the two sides perpendicular to it in tapering height upwards. Also, the height of bunds should be such that there was a free board of 25cm. above the maximum level of water standing in the rice field and the pond/tank. The efficiency of rice-fish farming system can be further increased by adding azolla (*Azolia microphylia*) to rice fields, i.e. by growing azolla in rice field throughout the cropping period as bio-fertilizer. Azolla not only adds to nitrogen availability to rice crop but also improves the fish food organisms. The fish excreta in turn fertilises the water in rice field.

While rice crop was grown in the catchment area of the pond during *kharif*, fish culture was undertaken in the pond with fingerlings of catla, rohu, mrigal and common carp. Care has to be taken to avoid the use of pesticides as a plant protection measure for rice crop and integrated pest management should be adopted instead. Neem-based pesticides, carbonates and low doses of selective organophosphates must be used so that fishes are not harmed. When water level in rice field recedes, the fish continues to live and grow in farm pond. The quantity of fish produced varied from 700 to 1000 kg/ha/year, and the net income from rice-fish farming was more than 30 to 50 percent over rice crop alone.

6.17.2. Fish-aquatic crop system

In flood prone areas, integration of fish farming with aquatic crops such as *makhana* (*Euryale ferox*) and water chestnut (*Trapa bispinosa*) has been found to be very remunerative and suggested as a tool for improvement of livelihood, particularly for small and marginal farmers. The system has been adopted by farmers in Darbhanga district of Bihar (NAIP 2012).

6.17.3. Rice-Fish farming system

Rice fields serving as a catchment to farm-pond in humid and sub-humid regions provide a perfect setting for rice-cum- fish culture. The bunds of farm pond/tank are strengthened such that 5 to 10 cms deep water was standing in rice field throughout the crop duration. The water was drained to pond completely 15 days before the harvest of rice-crop. The research work conducted at Fisheries Research Institute has shown that fresh water fishes can be cultivated in rice field-pond system. After the rice crop was harvested the fishes can retreat and confine themselves to the farm-pond. For protecting rice crop against insect-pest, integrated pest management must be adopted including neem-based pesticides, carbonates and selective organophosphates in low doses, so that fishes are not affected adversely. Vast tracts in eastern U.P., Bihar, Chhattisgarh and eastern Madhya Pradesh are very suitable for rice-cum-fish culture. A fish-yield of about 1000 kg/ha can be obtained resulting in at least 50 percent increase in income over mono-cropping of rice.

6.17.3.1. Species of fish: Catla and grass carp (*Cyernus carpia*) are fast growing species fish feeding on aquatic plants, cut grass and decomposing higher plants. Common carp (*Cyprinus carpio*) is omnivorous. Rohu (*Labeo rohita*) feeds on growing plants, decomposing vegetation, algae, zooplankton, etc. Mrigal (*Cirrhimus mrigale*) feeds on detritus, diatoms, algae, etc. It is advisable to go for polyfish or composite culture in farm ponds/tanks. Fingerlings of Catla, Rohu and Mrigal can be combined with those of common carp.

6.17.3.2. Management: In was necessary to nourish fish with supplementary feeds like rice bran and oilseed cakes to ensure faster growth and higher yield.

6.18. MORE INTENSIVE POND-BASED INTEGRATED FARMING SYSTEMS

For small and marginal farmers, there are numerous opportunities to integrate rice-fish farming with other livestock farmings, e.g. rice-fish cum duck farming, rice-fish cum poultry farming and rice-fish cum pig farming.

6.18.1. Rice-Fish cum Duck farming

The integration of duck farming with rice-fish culture was economically very advantageous to farmers, particularly the small and marginal ones. In this integrated farming system, production of both fish and duck can be increased while reducing the cost of fish-culture considerably. A series of experiments have been conducted to evolve a scientifically sound and economically viable system to integrate fresh water aquaculture with duck farming so that they benefit each other. The duck-house was so situated that droppings of ducks reach the farm-pond automatically, and manure it. Thus, droppings of the ducks are recycled to serve as feed to fish. About 250 ducks provide sufficient quantity of manure (12 tonnes ha^{-1} $year^{-1}$) to fertilize one ha of water surface area.

6.18.2. Rice- Fish cum Pig farming

The pig-house can be constructed on the bank of farm pond such that pig manure (excreta and urine) was directly drained into the pond. The pig excreta was directly utilised by fish as its feed because the excreta contained about 70 percent digestible food for the fish. Thus in this system of integrated farming, no supplementary feed for fish was required and accordingly no additional pond fertilization for fish-feeding was required. Obviously the cost of fish feeding and maintenance was reduced to about 60 percent. About 35 pigs are sufficient to provide adequate quantity of excreta required for manuring a hectare of water area of the fish pond. Under these circumstances a high fish-yield of 6000-7000 kg ha^{-1} $year^{-1}$ was generally obtained. The cost-benefit ratio of this integrated farming system was 1.5 to 2.0 with different pond sizes.

6.18.3. Rice-Fish Cum Poultry Integrated Farming System

Lowland or upland rice crop was planted according to situation and run-off got collected into farm pond/tank. The whole recommended dose of fertilizer should be applied as basal, and no split application of even nitrogenous fertilizer should be practised as it could depress fish growth due to high concentration of ammonia. Similarly as plant protection measure, no chemical insecticide should be used, as it could adversely affect the survival of fingerlings. Neem kernel extract (5 percent) could be used as plant protection measure. Composite fingerlings of fish species like grass carp, common carp, rohu and catla were reared as fish polyculture. The number of fingerlings released into tank/farm pond should be @10,000 per ha of rice crop.

For poultry rearing, poultry shed could be constructed right above the pond. It should be supported on concrete poles on all the four sides. The size of the shed should be 2m x 1.5m x 1m to accommodate 25 birds. The bottom of the shed should be of strong wire with 1cm^2 mesh so as to allow free fall of poultry droppings. The poultry dropping added every day serves both the purposes of fish-feed and manure to crop through pond-water. It has been observed that integrating fish and poultry with rice helps controlling weeds in rice-crop by 40 percent and pest incidence to a tune of 30 percent, as a bio-control measure. A net profit of more than one lakh rupees per year has been reported under farmers' conditions and thus this integrated farming system was very suitable for small and marginal farmers (NAIP, 2012).

Thus, there was tremendous scope not only to enhance the productivity of various rainfed crops but also to increase the intensity of cropping by recycling the collected run-off water. Further, by adopting the pond/tank based farming systems, an ever-green revolution under rainfed conditions in sub-humid and humid regions, at least of Central India, can be brought about.

■■■■

VII

CROPPING SYSTEMS AND MANAGEMENT PRACTICES

The appropriate cropping systems and their management practices should aim at the best possible utilization of land and water resources on a sustained basis. As already brought out in chapter II, the cropping systems being followed by most farmers under rainfed farming were irrespective of the type of land and soil they had and thus quite inappropriate. The research work conducted under the All India Coordinated Research on Dryland Agriculture at different centres had shown that (i) keeping the land fallow in *kharif* was most undesirable, and (ii) as rainfed crops both wheat and rice were the least remunerative crops, on a long term basis. Hence in Central India, these crops should be grown only on flat lands with deep and heavy soils, preferably in high rainfall areas or where at least one supplementary irrigation was available. Also, the practice of *kharif* fallowing must be done away with, particularly on black soil. The research work already referred to and the work done on farmer's fields under the Indo-UK Dry farming project had shown that soybean, short duration sorghum, maize and pigeonpea (short duration variety) crops on clay loam soils and groundnut and pigeonpea on loam and sandy loam deep soils were much more remunerative under rainfed conditions and there was an immense potential to increase their yields by right choice of proper varieties, timely sowing and adopting other improved practices of crop production.

7.1. UNTAPPED YIELD RESERVOIR OF DRYLAND CROPS

Although the average yields of some rainfed crops in Madhya Pradesh are higher than the national averages, but these are terribly low. The yield of soybean crop was comparatively higher particularly on black soils and the area under this crop increased by leaps and bounds to the extent that Madhya Pradesh state became famous as soybean state. Similarly in pulse production undivided Madhya Pradesh was holding quite high rank since 1973-74, and had been a surplus state in this respect. As a matter of fact, pulse crops were doing much better under rainfed conditions in the state and it was much more economical to the cultivators to grow pulse crop under rainfed conditions than to grow wheat after *kharif* fallow. There was a big gap between average yield of a crop for the whole State and the yield obtained on farmer's fields under Operational Research

Projects/National Demonstrations, through transfer of technology. Further, there was a gap between these yields (under ORPs, *etc*) and the yields obtained in research plots. The great potential of various upland crops was successfully demonstrated on farmer's fields in Indo-UK Dry Farming Project area (Table 38).

Table 38. Untapped yield reservoir in dryland crops

S.No.	Crop	Average yield (q ha^{-1})		Research plots
		Whole state	Indo-UK project and pilot project area	
1	Sorghum	7.4	23.0	30.0
2	Maize	11.1	27.5	35.0
3	Soybean	5.0	12.2	20.0
4	Groundnut	6.6	12.0	15.0
5	Wheat	7.1	9.6	12.5
6	Chickpea	6.0	7.7	15.0

7.2. APPROPRIATE CROPPING SYSTEM

There was an urgent need to adopt appropriate cropping system taking into consideration the type of land, depth of soil and its texture which determined the moisture holding capacity of the soil and the rainfall distribution in different parts of the central States. Within the same agro-climatic zone, the land slope, the type of soil and the depth of soil determined not only whether soil erosion or waterlogging was the problem but also the duration of moisture availability period. The length of this period for a particular locality could be worked out by taking into account the climatic elements *viz.* the daily rainfall and actual evaporation (worked out from daily potential evaporation data if the data on actual evaporation not available) and the soil moisture storage capacity. The soil moisture storage capacity for root zone of different soils was given in Table 39.

Table 39. The soil moisture storage capacity of root zone of different soils

S.No.	Soil Type	Moisture storage capacity (mm)
1.	Skeletal soils, yellow and red sandy soils	75
2.	Shallow black soils	150
3.	Medium black soil	200
4.	Deep black soils	300

7.2.1. Moisture Availability Period

Using the daily rainfall (P) and potential evaporation (PE) data available at College of Agriculture, Indore (M.P.) for the period 1931-70, moisture availability period was estimated following two approaches, viz. (i) simple soil water budget, and (ii) falling actual evaporation (AE) method in which

$$\text{Actual evaporation} = \text{PE} \times \frac{\text{Actual profile stored moisture on a particular day}}{\text{Moisture storage capacty of soil}}$$

In simple water budget, PE values are balanced against rainfall and profile stored moisture while in falling AE method, PE is multiplied by a factor derived by dividing the actual profile stored moisture by moisture storage capacity of the soil. The profile stored moisture values accepted for different depths of soils are 75, 150 and 300mm for shallow, medium and deep black soils, respectively, at Indore. The moisture availability periods calculated using both these approaches are presented in Table 40 (Verma and Robertson 1973).

Table 40. Maximum moisture availability period at Indore (at 80 percent probability)

Soil type	Moisture storage capacity (mm)	Moisture availability period (days)			
		Simple water budget		Falling AE method	
		Dates	Days	Dates	Days
Shallow black	75	July 8 to Oct. 6	90	July 8 to Oct. 26	110
Medium black	150	July 1 to Oct. 28	120	July 1 to Dec. 27	180
Deep black	300	July 1 to Dec. 27	180	July 1 to Feb. 3	220

The maximum moisture availability period worked out for other centres of All India Coordinated Research Project on Dryland Agriculture (AICRPDA) in Central India was shown in Table 41.

Table 41. Maximum moisture availability period in central India (ICAR, 1977)

Centre	Soil type	Mean annual rainfall (mm)	Climate	Moisture availability period (days)
Agra	Inceptisol	665	Semi-arid	170
Gwalior	Inceptisol	764	Semi-arid	160
Jhansi	Inceptisol	870	Semi-arid	150
Faizabad	Inceptisol	1051	Sub-humid	230
Jagadalpur	Inceptisol	1297	Sub-humid	300
Ranchi	Inceptisol	1149	Sub-humid	270
Rewa	Vertisol	1088	Sub-humid	250
Varansi	Inceptisol	1049	Sub-humid	230

As shown in Table 41, moisture availability period in central India varied from 150 days to 300 days – a big range. It was the least for Jhansi (Bundelkhand region) and the highest for Jagdalpur (Bastar Plateau).

7.2.2. Appropriate crop planning according to land suitability and moisture availability

The duration of crops/varieties to be grown on different soil types had to be tailored to the moisture availability period. Also suitable silvi-pastoral, horti-pastoral, agri-horticultural and agro-forestry systems should be adopted and appropriate plant species and crop plants have to be grown depending upon the type of land and soil and the extent of erosion. While quick growing trees for fuel, fodder and fruits had to be planted on hill slopes, grasses must be grown on slopy land with shallow soil. If the soil was shallow but the slope was less than three percent, *kharif* fodder crops or short duration varieties of sorghum and soybean and short duration *kharif* pulse crops were most suitable. Short duration varieties of sorghum and soybean gave high yields even in drought years of 1974 and 1979. Also these gave good yields even on shallow soils where long duration varieties of sorghum (desi, pilliaple, Vidisha 60-1 etc.) and soybean (Kali tur and T 49) failed to produce any seed (Indo-UK DFP 1980b). Recommendations on crop planning according to suitability of land are presented in Table 42.

Table 42. Crops for various land/soil conditions

S. No.	Land/soil conditions	Suitable trees and crops
1	Steep hill slope	Fruit trees such as custard apple and aonla, fuel trees such as eucalyptus, unjal (*Albizzia amara*), *siris* (*Albizzia lebbek*), *babool* (*Acacia arabica* and *A. tortilis*).
2	Sub-hill region and common pasture lands	Grasses such as *dinanath* (*Pennisetum pedicellaum*), marvel (*Dicanthium*), pohna (*Sehima sulcatum*), chipkni (*Desmodium contortum*), tore (*Thalipogon aligans*), kel (*Bothriochloa*) with legumes such as siratro (*Phaseolus atropurpureous*) and pilli pasara (*Phaseolus trilobus*) and sheora (*Alysicarpus rugosus*), fodder shrubs such as subabul (*Leucaena leucocephala*).
3	Shallow black soil	Sorghum (CSH5, CSH 6, CSH 9) maize (Ganga 5, Chandan 3), soybean (JS 72-44, JS 71-5).
4	Medium black soil	Soybean/sorghum + pigeonpea (ICPL-87, TJT-501) soybean-cotton, maize-soybean, sorghum for fodder (Vidisha 60-1), followed by safflower or gram.
5	Deep black soil	Maize - safflower (JSF 1), soybean-chickpea (JG-16), sorghum-chickpea, maize-chickpea, soybean-safflower, and soybean-wheat (with come up irrigation, or on receipt of late rain).

7.2.3. Cropping in *kharif* a must

Growing a crop during rainy season not only makes the best use of rain-water but also serves as the cheapest and effective soil conservation measure. As a matter of fact, if the land has a slope of 1.0 to 3.0 percent, the soil erosion due to run-off water on such land can be fully controlled by raising a well nourished crop, planted against the land slope on a grade of 0.2 to 0.3 percent. If improved practices of growing the *kharif* crop are adopted, the crop-cover provided by it will be better and more effective. Cropping in *kharif* selecting high yielding varieties and using all the improved package of practices, e.g. selection of good quality seed, seed treatment, fertilizer dose based on soil-test, intercultural operations, weeding and plant protection measures not only resulted in high yields and the consequent high income to the farmer but also protected his land and ensured stable farming system. Fortunately, in *kharif* season soil moisture was not a limiting factor normally and therefore fertilizer and manure application could be adopted profitably unlike in *rabi* season.

7.2.4. More remunerative crops

For single cropping in *kharif* or *rabi* season, following crops have been found to be more remunerative.

7.2.4.1. Soybean: As early as 1972-73, the All India Coordinated Research Project on Dryland Agriculture, main Centre College of Agriculture, Indore (M.P.) established that soybean was the most remunerative *kharif* crop on black soils. Through Indo-UK Dry Farming Project, Indore, this finding was tested and transferred to farmers' fields in the project area (Nainod, Jamburdi Hapsi and Rinjlai villages of Indore tehsil) and soon the crop spread not only over whole project area but also far and wide on black soils. At Ranchi Centre, maize and sorghum were found to be more remunerative than upland rice.

7.2.4.2 Safflower: For *rabi* season, safflower crop proved more remunerative as compared to already popular crops (wheat, chickpea, linseed etc.) grown under rainfed conditions in Malwa. But it could not be adopted by farmers unless marketing facilities and a minimum support price by the Government were ensured. At Agra Centre mustard crop was better alternative to wheat, while at Varanasi Centre chickpea proved better than wheat.

7.2.4.3. Fodder crops: Over large tracts, left fallow in *kharif*, particularly under black soil, fodder crop of sorghum and teosinte (makchari) with and without cowpea could be raised not only to provide protection against soil-erosion but also to provide fodder for livestock without interfering with raising a good *rabi* crop even in a year during which monsoon ceased early.

7.2.5. Cropping systems according to moisture availability

Traditional cropping system in a region depended mainly on its climate, soil, and food-habits of inhabitants. It may not be the most remunerative in terms of productivity per unit area and per unit amount of available moisture. Since soil moisture is usually the most limiting factor for growth and development of crops under rainfed condition, particularly in semi-arid areas, raising a crop during monsoon season was safer and more productive and thus more profitable than growing crop on profile stored moisture. Further, intensity of cropping could be increased only by cropping in *kharif* instead of leaving the land fallow in *kharif*. If distribution of rainfall was favourable, even in semi-arid region with Vertisol, double cropping could be raised. More remunerative and reliable cropping systems for different moisture availability periods were presented in Table 43.

7.2.6. Double cropping under rainfed condition

As shown in Table 43, double cropping under rainfed condition was assured when the moisture availability period was greater than 250days in a year. In middle Gangetetic plain with sub-humid climate and deep alluvial soils, blackgram, sesame, clusterbean, cowpea and groundnut could be grown in *kharif* followed by chickpea, linseed, sunflower and safflower in *rabi*. In Chhattisgarh and middle Gangetic plains and northern Jharkhand state, rice was grown on large tracts and this could be followed by *utera* crop of lathyrus (toxin free varieties), chickpea and mothbean. *Utera* was the practice of broadcasting seeds of lathyrus etc. in standing crop of rice (Relay cropping). Alternatively, the seeds of chickpea or linseed were drilled between rice crop stubbles immediately after the harvest of rice crop. In southern Jharkhand, maize and groundnut crops in *kharif* could be followed by linseed, lentil and barley in *rabi*. Over large areas under Vertisols and associated soils, the moisture availability period (80 percent probability basis) was more than 250 days, and the double cropping was almost assured. Further, there was always sufficient profile stored moisture in these soils in lower layers (below 15cms from soil surface) as shown in table 44. *Kharif* crops upto time of their harvest depleted moisture in surface soil only and this moisture was critical for germination of seeds of second crop to be grown in *rabi*. This situation (less than 24 percent) arose only when either the monsoon receded early (before September 15th) or the crop/variety grown in *kharif* was of long duration. Also, it was not the total rainfall received during *kharif*, but its distribution, which was important (Sharma and Thukral 1987).

Table 43. More remunerative cropping system for different moisture availability periods

Moisture availability period (days)	Soil type	Suitable cropping system
150-180	Alluvial soil, Mixed red yellow soil and Shallow balck soil	Single *kharif* crop (sorghum, sesame, maize, pearlmillet, blackgram, greengram, clusterbean)
	Deep black soil	Single *kharif* or *rabi* crop (soybean, safflower)
180-250	Alluvial soil	Intercropping (sorghum + pigeonpea, pearlmillet + pigeonpea)
	Mixed red and yellow deep soil	Sole rice crop
	Medium and deep black soil	Intercropping (soybean + pigeonpea)
>250	Alluvial soil	Blackgram – maize/pearlmillet Maize/sorghum – chickpea Blackgram – mustard Rice – chickpea
	Mixed red and yellow soil Deep black soil	Maize – chickpea Soybean – chickpea Rice – chickpea

7.2.7. Intercropping systems

Where double cropping could not be practiced under rainfed farming, the cropping intensity could be increased by adopting appropriate intercropping system. In Bundelkhand, Malwa and Vindhya regions, soybean + pigeonpea (4:2 rows) intercropping was most paying. On light textured but deep yellow, red and brown soils (Alfisofs) sorghum/pearlmillet + pigeonpea (2:1 or 4:2 rows), blackgram + pigeonpea, or groundnut + pigeonpea (4:2 rows) were recommended. On black soils, maize + soybean (2:2 rows) and sorghum + pigeonpea (2:1 or 4:2 rows) intercropping systems were also recommended besides soybean - pigeonpea intercropping. In Vindhya Region, mixed cropping (mixing seed of two or more crops and broadcasting the mixed seeds) was very common and it was successfully replaced by intercropping systems and the later were found to be more remunerative (Sharma 2004).

Table 44. Moisture content (volumetric) in Vertisol at the time of harvest of *kharif* crops in different years

Soil layers (cm)	Years and crops grown in *kharif*									
	1980-81		1981-82		1982-83		1983-84		1984-85	
	Soybean	Maize	Soybean	Maize	Soybean	Maize	Soybean	Maize	Soybean	Maize
0-15	16.8	17.2	19.3	21.4	19.3	21.3	32.5	30.1	15.5	17.6
15-30	20.1	20.5	27.8	27.7	27.9	24.1	31.8	30.8	19.3	19.7
30-120	29.5	30.7	32.0	34.6	35.6	32.9	38.8	36.7	30.5	31.6
Whole profile (cm.)	32.2	33.8	35.7	38.8	39.1	36.8	44.6	42.1	34.9	34.3

7.3. IMPROVED PACKAGE OF PRACTICES

It is necessary to adopt improved package of practices for not only getting higher net return but that too on sustainable basis.

7.3.1. Conservation practices

Certain cultural practices recommended for higher and stable yields of different crops are given below.

1. Sowing crops against slope in slopy land.
2. Planting on grade (about 0.3 percent) rather than on graded broad ridge, not only facilitated inter-culture for weed-control but also reduced soil-erosion.
3. Providing surface drains/graded furrows on flat land (between crop-rows) or planting on ridge and furrow system under heavy rainfall and dark clay soil conditions.

The last two recommendations were found to be better substitute for broad bed and furrow system (BBF).

7.3.2. Selection of varieties

In rainfed farming, short duration crops and early varieties of various crops were ultimately more paying over time, as these escaped drought. Not only this, these vacated the fields early and therefore, chances of double cropping under rainfed conditions improved in case there was a late rain or just one irrigation could be managed from well or by run-off collection and recycling. Further it was easier to replace varieties of already popular crops than to replace old crops by new ones. Such varieties of different crops evolved and recommended by J.N.K.V.V., Jabalpur and I.G.K.V.V., Raipur were shown in Table 45.

Table 45. Recommended crop varieties for rainfed farming areas

Crop	Malwa region (deep soils with medium rainfall)	Rewa Region (medium light textured soils with high rainfall)
Maize	JM 8, HM 12, JM 216, HBM 421, Ganga safed 2,	Deccan, HBM-421, HPOM-1
Sorghum	CSH 18, Indore 12,JJ 938, JJ 1022, JJ 741	CSH 5, Swarna, Jawahar 236
Soybean	JS 95-60, Ahilya 3 (NRC-7), Durga (JS 72-280), JS 97-52	JS 75-46, JS 76-205, NRC-12, JS 70-280
Groundnut	JL-24, TJM 3, TG 26, TG 41	JM 11, JM 13, TJ 37, TGN-23,
Redgram (Pigeonpea)	JA 4, KMT 7, TJT 501, Khargaon 2, ICPL 89	Prabhat, JKM 189, JA 3, RVA 28,
Cotton	Khandwa 2, RB 50, Badnawari, Vikram, K 4	-
Chickpea	JG 16, JG 412, JG 11, JG 315, JG 74, JG 5 (Pink)	JG 221, JG 74, Gwalior 2, JG 53 (pink) Narsinghpur bold
Wheat	Sujata, JW 517, HW 2004, H 91500, Narmada 112, Narmada 195	C 306, Narmada -112
Sunflower	JSN 1, MSFH 17	JSN 1, MSFH 17
Linseed	JLT 27, JLS 9, JLS 23-10, JLT 26	JLT 27, JL 23
Safflower	JSF 1, JSF 7, JSF 73, RVS 113	JSF1, JSF 7, JSF 97
Paddy	-	JR 201, JR 353, JR 345, JR 504 (Upland) JR 64, Madhuri, (Lowland)
Sesamum	-	JT 21, JT 55, Kanchantil (JT 7), TKG 8, TKG 308
Black gram	JU 3, PDU 1, LBG 20, TAU 1, T 9, PU 39	T 9, Mash 48, Rewa 2, Gwalior 2, Jawahar Urd 3
Green gram	JM 721, TARM 1, LGG 460, PDM 139, PDM –11	JM 721, TARM 1, PDM 139
Kodo	-	Niwasi, IPS 236-1, Jawahar kodo 2, JNK 62, JNK 41,
Kutki	-	Dindori-1, Jawahar Kutki 2, Jawahar Kutki 8
Lentil	JL 3, JL 1, IPL 81, RVL 31	JL 3, JL 1, IPL 81

7.3.3. Other improved practices

It is necessary to adopt following improved practices for higher net return.

1. Appropriate but not excessive tillage for preparation of (weed-free) good seed bed.
2. Seed grading/treatment with fungicides and of legumes with culture.
3. Early but safe planting (dry sowing of only cotton).

4. Line sowing of *kharif* crops against traditional method of broad casting seed.
5. Seed and fertilizers should be drilled using ferti-seed drill so that both the seed and the fertilizer could be placed in moist soil, and seed was placed above fertilizer.
6. Integrated nutrient management through judicious use of fertilizers/ biofertilizers and organic manures.
7. Control of weeds by intercultural operations or by hand. If the field conditions did not permit such operations, use of appropriate weedicides was recommended.
8. Control of pests and diseases using integrated pest management.
9. Harvest of crops at physiological maturity (to get vacated fields early and avoid losses during harvest) and
10. Adoption of post-harvest technology.

7.3.4. Practices needing more emphasis

Date of planting, appropriate plant population, and integrated nutrient management under rainfed farming needed more emphasis and were therefore, discussed below.

7.3.4.1. Early but safe sowing: Sowing of both *kharif* and *rabi* crops as early as possible gave high yields. However, it could not be done too early for moisture factor in *kharif* and temperature consideration in *rabi*, particularly for wheat. Farmers can start with dry sowing of cotton in the week with 80 percent probability of break of monsoon. Sowing of sorghum, maize and soybean can be taken up with onset of monsoon, followed by groundnut and pigeonpea (in case of sole crops, as intercrop it would be sown with companion crop). Sunflower did better as late *kharif* crop while rapeseed, toria and safflower performed best when sown quite early as *rabi* crops. Linseed and gram could be sown as normal *rabi* crops followed by wheat (when temperature had come down). For determining appropriate time of sowing of *kharif* crops, moisture situation in soil, probability of rainfall, etc. had to be considered and this might vary slightly from year to year depending upon amount and starting date of pre-monsoon showers, gap between pre-monsoon showers and on-set of monsoon, etc. Early sown sorghum escaped shoot fly attack.

7.3.4.2. Appropriate plant population: Adequate plant population was the most important problem in increasing yields of rainfed crops. While date and method of planting affected the plant population, quality of seed was equally important. Before sowing, germination percentage of the seed should be determined and seed-rate increased according to germination count.

7.3.4.3. Integrated Nutrient Management in rainfed farming: Undoubtedly, the drylands are not only thirsty but hungry as well. Vertisol and associated soils (Deep, medium and shallow black soils), Alfisol and associated soils (Red, Mixed red and yellow soil, Mixed red and black soils) Inceptisols and Entisols (Alluvial soils) were the main soil types covering Central India. The soils were in general low in organic carbon and nitrogen content, low to medium in phosphorus and medium to high in potassium. But due to continuous mining of soil K, its status was also changing to low to medium. Sulphur status was low for oilseed and pulse crops. Deficiency of iron and zinc was also emerging in some areas. Still average fertilizer consumption in rainfed areas was about 25kg ha^{-1} against the national average of 90kg ($N + P_20_5 + K_20$) ha^{-1}. Only 10 percent (1.8 million tonnes) of total fertilizer nutrients used in the country are applied to rainfed areas against the removal of 7.4 million tonnes by rainfed crops. Thus, 5.6 million tonnes of nutrients are being mined from the soil which already had a poor fertility status. The fertilizer response ratio of various rainfed crops under frontline demonstrations conducted at different centres of Soil Test Fertilizer Response Correlation scheme varied from as low as 4.0 to 17.0kg grain kg^{-1} nutrient (Biswas and Sharma 2005). The response was greater by *kharif* (rainy season) crops as compared to *rabi* (winter season) crops. Also, it was generally greater on Vertisols and associated soils as compared to Alfisols.

A number of studies on benefit : cost ratio of fertilizer use under rainfed conditions showed that at 15-80kg N ha^{-1}, the ratio was 3.7 to 7.7 for sorghum, 4 to 5 for pearlmillet, 2.6 to 4.2 for rice and 5.7 to 13.4 for maize. At the highest level of 60kg N ha^{-1}, the benefit : cost ratio was 3.2 to 7.1 for mustard and 7.4 for safflower (Chowdhary, 1979).

For *kharif* crops, particularly if grown on black soils, moisture was usually not a limiting factor for crop-growth and therefore, very good results were obtained on fertilizer use. For *rabi* crops, the dose of nitrogen and phosphorus should be varied according to the quantity of profile stored moisture and the amount of winter rainfall received. Fertilizer trials on rainfed crops conducted in different agro-climatic regions had shown that-

i. The yield of different crops increased when recommended doses of fertilizer based on soil tests were used. There was huge literature reporting significant response of different crops.

ii. Yield increase was more when both nitrogen and phosphorus were applied, proving the importance of balanced fertilizer-use. While response to N was universal under good growth conditions, significant response to P or even K has now been obtained.

iii. In case of *kharif* crops, half the dose of nitrogen and full dose of phosphorus and potassium (if necessary) must be applied at the time of sowing, preferably by drilling below the seed. Remaining half dose of N should be applied as top-dressed before intercultural operations in *kharif* crops. But if the crop was to be grown on profile-stored moisture basal application of all nutrients was recommended.

iv. An integrated nutrient supply plan must be adopted to provide the recommended doses of various plant nutrients, so that resources available with the farmers e.g. compost, farm yard manure (FYM) etc. were also utilized with comparatively less expensive bio-fertilizers (Azotobactor, Rhizobium culture and other inoculants like Phosphate solubilising bacteria, etc.) to supplement the costly fertilizers. Use of compost and FYM had shown better results under rainfed farming. Also, efficiency of fertilizer nutrients was increased when fertilizer was used with organic manure, i.e. there was synergistic effect when used together. Organic manures supplied various essential nutrients (though in small amounts) and were therefore more balanced as compared to fertilizers. Also, these improved physical condition of the soil required for favourable soil-water plant relationship. But the content of major nutrients viz. N, P and K were so low that crop requirement of these could not be met fully by organic manures alone and it had to be supplemented by fertilizers. Use of effective rhizobium cultures to all leguminous crops was not only cheaper but also quite useful, particularly in soybean crop whose demand for nitrogen was enormous (Vittal 1987).

v. **Other nutrients also important**: Deficiency of nutrients other than N and P was also met with in certain areas. With continuous use of urea and diammonium phosphate (DAP) for long, the deficiency symptoms of sulphur were observed on mustard and soybean crop. To control this deficiency, use of DAP must be substituted by superphosphate or sulphur must be added in suitable form. Similarly, if deficiency of zinc was observed in some pockets, zinc sulphate @ 25 kg ha^{-1} must be applied at the time of sowing.

7.3.4.4. Integrated management: Integrated management of land, water and plant-nutrients had synergic effect in increasing the productivity of rainfed farming. The inputs of soil and water conservation, improved seeds and fertilizer had greater impact on crop-yields compared to traditional practice as shown in table 46 (Biswas and Sharma 2005).

Table 46. Correlation of various inputs on yield-increase

S.No.	Input	Percent yield increase over traditional practices
1	Management practices	14
2	Seed	40
3	Fertilizer	50
4	Seed + Fertilizer	90
5	Seed + Fertilizer + Management practices	130

Fertilizer use under rainfed conditions results in more efficient utilization of profile stored moisture and under certain circumstances helps evade adverse effects of drought as crop is able to use water from deeper soil layers.

7.4. RECOMMENDATIONS FOR DIFFERENT AGRO-CLIMATIC REGIONS

Certain specific recommendations on cropping systems had emerged from research work conducted at Dryland Research Centres, Agra, Varanasi, Ranchi, Jhansi, Indore and Rewa and at various Regional Research Stations of JNKVV,

Jabalpur and IGKVV, Raipur (AICRPDA, 1983, CRIDA, 1987), and the same were presented below. These recommendations were made assuming that problems of waterlogging, if any, had been taken care of.

7.4.1. Gird region and Upper Gangetic plain (south west U. P.)

Under rainfed condition, rainfed mustard and chickpea were found to be more remunerative than rainfed wheat. Blackgram – mustard crop sequence under unirrigated and limited irrigation conditions proved more remunerative in Gird region. Sorghum hybrid + pigeonpea and pealmillet + clusterbaen were better intercropping systems in Gird region. Pearlmillet + greengram (2:2) inter cropping was more remunerative at Agra. High yielding varieties of groundnut, mustard and pearlmillet needed to be spread widely. More than 30percent cropped area in upper Gangetic plain was irrigated by canal and tube-wells and rice, wheat, sugarcane, vegetable, etc. were grown.

7.4.2. Malwa region

Fallow-wheat must be replaced by soybean-fallow. Soybean + pigeonpea inter cropping was more remunerative than soybean sole crop. Also depending upon rainfall distribution, maize, hybrid sorghum, and soybean could be followed by safflower and chickpea crops in *rabi*. Safflower performed better at Indore and southwards and not northwards (at Varanasi and Dehradun), as it was more susceptible to frost compared to wheat (CRIDA, 1987).

7.4.3. Kymore plateau

On shallow soils and slopy land, wheat and paddy crops must be replaced by pigeonpea, soybean and groundnut. Of other crops, blackgram and safflower proved more remunerative than sorghum and sesame on flat land with deep soil. Also, soybean was more remunerative than rainfed rice on flat land. Sorghum – chickpea crop sequence and sorghum + pigeonpea intercropping were recommended.

7.4.4. Vindhyan Plateau

Greengram and blackgram must be spread on upland while soybean (short duration) should be spread widely on flat land with deep black soil. With run-off collection, soybean-wheat on flat land and groundnut-wheat on well-drained land was recommended.

7.4.5. Northern Chhattisgarh Hills and Chhotanagpur plateau of Jharkhand

On shallow soil and slopy land, hybrid sorghum and high yielding varieties of millets must be introduced. Also, upland rice must be replaced by maize, sorghum, fingermillet and groundnut where rainless periods of a week or more, drastically reduce the yield of upland rice. Pigeonpea must be used as an intercrop (CRIDA, 1987). On flat land, soybean was more remunerative than rice, and rapeseed and mustard should be popularized in *rabi*. Also, horticulture development must be intensified.

7.4.6. Satpura plateau

Groundnut, maize, minor millets and vegetables (*kharif*) on light and shallow soil should be introduced. Soybean on heavy soil was recommended to replace rice. With run-off collection soybean-wheat crop sequence was more remunerative while cotton or sole crop of soybean or cotton-soybean intercropping was recommended under conditions where run-off collection was not done. In northern parts of Chhindwara and Seoni districts, soybean and pigeonpea inter cropping must be introduced on flat land with black soil (DES 1979).

7.4.7. Nimar plains

Sorghum on shallow soil, groundnut and pigeonpea on medium soil, and cotton alone or inter cropping by soybean on deep soil proved more remunerative.

7.4.8. Narmada valley

At present haveli system was practiced which made the least efficient use of rain-water. Soybean in *kharif* after providing external drainage and wheat in *rabi* after run-off collection and recycling would usher in green revolution in rainfed areas of this region.

7.4.9. Jhabua Hills

Maize and paddy crop on light soils must be replaced by millets and sorghum. In valleys, soybean or cotton should be grown in *kharif* and safflower in *rabi*.

7.4.10. Bundelkhand Region

Bundelkhand region was one of the most underdeveloped region of the country due to inappropriate management of land and water resources resulting in frequent crop failures. On well drained land, intercropping of pearlmillet and pigeonpea, of sorghum and pigeonpea, of groundnut and pigeonpea (2:1) and blackgram-mustard crop sequence were more remunerative while on flat land with medium and deep black soil, intercropping of soybean with pigeonpea, of sesame with pigeonpea or soybean alone was recommended. Also *kharif* pulse crop such as cowpea, greengram, blackgram could be followed by mustard, linseed, sunflower and chickpea in *rabi* (CRIDA, 1987). Two rows of sesame intercropped with two rows of sorghum was the most remunerative system on red soil. Uncultivated hillocks must be used for silvi-pastoral system. Singh (1996) indentified four broad ecological situations as follows:

a. Moderate to high rainfall (800-1000mm), heavy clay soils, flat to moderately slopy land, suitable for both *kharif* and *rabi* crops (Jalaun and part of Hamirpur and Banda districts). This was the most productive part of the zone.

b. Moderate rainfall (500-800mm), red and laterite soil, moderately slopy and well drained land, moderate to severely eroded, suitable for oilseeds, pulses and fodder and forage crops (Banda, Hamirpur, Jhansi and Lalitpur districts).

c. Moderate rainfall (500-800mm) red and laterite soils, undulating and severely eroded land suitable for agro-forestry and silvi-pastoral systems.

d. Moderate rainfall (500-800mm) extremely slopy land dominated by exposed rocks, red and lateritic soils in patches, suitable for plantation only after soil conservation measures.

7.4.11. Chhattisgarh plains or Mahanadi Plains

On *bhata* land, paddy should be replaced by kodo, sorghum, blackgram and niger. Alternatively, silvipastoral/hortipastoral systems should be developed on such land planting trees like aonla, lucaena etc. along with improved grasses. On unbunded *bharri* (black) soil intercropping of soybean+pigeonpea under unirrigated conditions and soybean-mustard crop-sequence under limited irrigation conditions, are recommended. Rapeseed and mustard crop should also be popularized. Linseed, lentil and chickpea can replace lathyrus as *utera* crop.

7.4.12. Bastar plateau

On well drained land, rice must be replaced by soybean. SRI (system of rice intensification) method of rice cultivation must be popularized for higher productivity. For rice based *utera* cropping system, fieldpea, chickpea or linseed must popularized as *utera* crop to replace the traditionally grown lathyrus crop.

7.4.13. Central Gangetic Plains of eastern U.P., Bihar and northern Jharkhand

There was a great scope to increase yield of traditional paddy-wheat cropping system by adoption of improved technology evolved by research.

Short duration rice varieties should be sown on midlands and low lands to allow second crop of safflower, chickpea, linseed, lentil and barley to be grown in *rabi* season. Blackgram-mustard, sesame-chickpea were suitable crop sequences. On upland maize was a good substitute for rice where it suffered from intermittent drought.

Region specific problems such as waterlogging, floods and improper drainage should be solved by appropriate management of rain-water as already discussed.

■■■■

VIII

PLANNING FOR ABERRANT WEATHER CONDITIONS AND DROUGHT MANAGEMENT

In Central India, no major irrigation project was launched during last twenty years and therefore, whatever increase in irrigated area had taken place in Bihar, Madhya Pradesh, Jharkhand and Uttar Pradesh states during last two decades was due to irrigation through tanks, tube-wells and open dug wells. The area of food grain crops irrigated (CGWB 2006) in Uttar Pradesh, Bihar, Madhya Pradesh, Chhattisgarh and Jharkhand states was 71.6, 59.3, 39.7, 24.3, and 4.3 percent, respectively, and looking at sources of irrigation in the areas, 82.7, 77.9, 72.1, 68.8 and 29.5 percent irrigation in M.P., U.P., Bihar, Jharkhand and Chhattisgarh, respectively, was through tanks, tube-wells, open dug wells and other sources which depended on rainfall directly or indirectly, for recharge. Thus, farming in these central states depended heavily on rainfall and was affected adversely by aberrant weather, as main source of water for rainfed crops was rain. The effects of green revolution vanished completely during a drought year as both plant and animal life suffered due to non-availability of water whenever and wherever needed. As a matter of fact 'normal rainfall' was only a statistical figure worked out as long term average and the rainfall in any individual year deviated from the normal because rainfall always varied in time and space. Further, the rainfall was highly seasonal as more than 80 percent of the total annual rainfall was received through southwest monsoon during June to September in the whole country except Kashmir, Tamilnadu and eastern part of the Peninsula. The farming in rainfed area was a gamble with monsoon as bursts of rainfall alternated with dry breaks throughout the rainy season. Such vagaries of monsoon made the weather erratic/aberrant. The aberrant weather-drought or excessive rainfall in different years was quite common. Excessive rainfall in a year though caught attention, did not cause as alarming a situation as created by drought. Excessive rainfall might result in failure of crops completely or partly but did not cause shortage of fodder and drinking water. Also, excessive rainfall in *kharif* season always created better moisture conditions for *rabi* crops and thus compensated for any loss in *kharif* season. That was why excessive rainfall was always welcome much more than drought.

8.1. ABERRANT WEATHER

Weather conditions vary considerably not only from year to year but also within the year. The rainfall distribution is not only erratic but uncertain and consequently, strategy of dry farming may have to be changed during a year so as to meet the aberrant weather. The weather may be aberrant on account of:

a. Too early onset of monsoon (before June 15) followed by big gap (up to 20 days) and early withdrawal (before October 2),
b. Too early onset of monsoon without big gap but early withdrawal.
c. Too early onset of monsoon without gap and late withdrawal (after October 2).
d. Normal onset (about June 20) without big gap but early withdrawal.
e. Normal onset with a big gap and late withdrawal.
f. Late onset of monsoon (after June 20) without big gap but early withdrawal, and
g. Late onset of monsoon with a big gap and late withdrawal.

The selection of crops and their sowing time will vary under aberrant weather conditions. Also, mid–season corrections may be necessary when the weather has been too aberrant for the crops chosen for a normal weather to start with. The salient features of refinement of crop planning for different aberrant weather conditions are discussed below.

8.2. EARLY ONSET OF MONSOON FOLLOWED BY A GAP AND EARLY WITHDRAWAL

Such a weather pattern means that soil will get saturated with moisture quite early in the monsoon season, and following dry period will enable the farmer to prepare seed-bed and sow the *kharif* crops. Short duration crop/varieties will be better under such conditions. Once the soil column has got saturated up to about 20 cm, the farmer should not hesitate sowing cotton, sorghum, pearl millet and groundnut crops as early as soil conditions permit. For such conditions, sorghum/pearlmillet was better than maize, and groundnut was better than soybean crop.

8.3. EARLY ONSET OF MONSOON WITHOUT GAP BUT EARLY WITHDRAWAL

The strategy will be same as under (a) above, except that more attention had to be paid to providing field drainage on deep black soils by opening furrows on 0.2-0.3% grade across the slope with the help of a local plough fitted with a wooden plank. The inter space between furrows should depend upon crop to be grown (less inter space for maize than sorghum and soybean) and depth of soil (less inter-space on deeper soil). The adoption of complete package of improved practices made crop-growth more rapid so as to smother the weeds. Also, adoption of 'year round tillage' for weed–control was particularly useful under such weather conditions as time available (due to soil conditions) for inter culture by bullock-drawn implements was invariably short. However, inter culture by hand operated tools like 'Jalo Cultivator' was much more efficient than hand-hoe (*khurpi*).

8.4. EARLY ONSET OF MONSOON WITHOUT BIG GAP BUT LATE WITHDRAWAL

Such a weather means excessively wet conditions although the season and thus provision of drainage particularly on deep black soil was a pre-requisite for the success of *kharif* crops. Weeds had to be controlled by hand-weeding or by the use of weedicides (chemical weed killers). However, double cropping would be assured if a short duration *kharif* crop/variety was grown. Also, sorghum (varieties CSH-18, JJ-1022) and soybean (varieties JS 95-60, Ahilya-3 (NRC-7), JS 97-52 and Durga (JS-72-280) would do better than maize which could be recommended only on shallow black soil or on light soils. The grain of these sorghum varieties was not discolored even under continuous rains at grain formation stage or later. Further, as these varieties got off the field early, second crop could be planted starting with safflower in September, and then chickpea and wheat in October. Higher yields of these *rabi* crops could be obtained by drilling slightly higher doses of fertilizers at the time of sowing to make better use of higher quantity of available moisture (as a result of late withdrawal of monsoon), e.g. 30 kg N + 30 kg P_2O_5 ha^{-1} could be used instead of 20 kg N + 20 kg P_2O_5 recommended to wheat and safflower in a normal season. For gram, 40 kg P_2O_5 ha^{-1} could be used instead of 25 kg P_2O_5 ha^{-1} for a normal season.

8.5. NORMAL ONSET WITHOUT GAP BUT EARLY WITHDRAWAL

The pre-monsoon showers must be used advantageously to get a weed-free seed-bed. Sorghum and maize (on well drained fields) will do better than long duration crop like cotton. Also short duration varieties of sorghum (CSH-18, JJ-1022) will out yield long duration variety e.g. Vidisha 60-1 or desi sorghums which suffer from moisture stress at grain formation stage due to early withdrawal of monsoon. Under these conditions, safflower after fallow or short duration *kharif* crop (depending upon soil moisture) will be better than wheat or gram grown after *kharif* fallow.

8.6. NORMAL ONSET WITH A BIG GAP AND LATE WITHDRAWAL

Under these conditions, crops planted in *kharif* might survive or fail depending upon the length of drought spell. Sorghum was more tolerant to drought than maize and soybean. If crops of cotton, maize and soybean fail completely or in patches, sowing of sunflower should be done as a mid-season correction measure to replace cotton, etc. or to fill in gaps in these crops if failure was in patches only. As regards sorghum, it could be harvested for fodder (if HCN content was not high) and ratooned for grain. The intercropping of sorghum with pigeon pea would be particularly remunerative under these conditions as higher yields of pigeon pea could more than compensate the low yield of sorghum.

8.7. LATE ONSET OF MONSOON WITHOUT BIG GAP BUT EARLY WITHDRAWAL

Short duration varieties of sorghum and maize and groundnut (on shallow soils) may do better than cotton and long duration varieties of sorghum and maize. Seed treatment with carbofuran must be done before sowing. Early planting of safflower can be done as relay cropping or by using 'till and plant machine' without any preparatory tillage after harvesting the *kharif* crops at physiological maturity stage. Collection of run-off water will enable growing a second crop of chickpea and wheat when used for come-up irrigation.

8.8. LATE ONSET OF MONSOON WITH A BIG GAP AND LATE WITHDRAWAL

Short duration varieties of sorghum and groundnut must be preferred for sowing in *kharif*. Treatment of seed with carbofuran must be done. Maize was not suitable as it was much more susceptible to drought. Adequate quantities of fertilizer drilled below seed at the time of sowing will be an insurance against the dry spell. Late withdrawal of monsoon enabled growing a second crop of safflower, chickpea or even wheat depending upon when the monsoon withdrew. Collection of run-off water and its recycling may provide life-saving irrigation to *kharif* crop and may make double cropping a surety due to higher soil moisture content at the time of harvest of *kharif* crop.

8.9. HOW TO START

As a result of experience of Dryland Research Project Centres Indore and Rewa in M.P. and others throughout the country, it was known what was the best crop plan under particular weather conditions. But unfortunately the weather was most unpredictable and therefore, farmers should start with practices having the least risk. For example, farmers should plan to grow various *kharif* crops in different fields (but their short duration varieties only), befitting his land (according to slope, soil depth and drainage) rather than a single crop. Also, the farmer should plan for normal weather taking some precautionary measures in advance, proceed according to suitability of prevalent weather conditions, and making mid-season corrections later on, if necessary.

8.10. ADVANCE ACTIONS AND PRECAUTIONARY MEASURES

Certain actions have to be taken before onset of monsoon to minimize the adverse effects of drought. These are:

8.10.1. Seed treatment

Under drought conditions, the seed was likely to be affected by soil-borne diseases resulting in poor germination and low plant population. Good quality seed from reliable source should be obtained. If farmers wanted to use their own seed, grading and treatment of the seed with fungicide and appropriate culture should be done to realize the full potential of the seed and assure optimum plant population.

8.10.2. Crop rotation

Following a legume (pulse)-cereal (sorghum etc.) crop rotation assured better yields than cereal – cereal system in a field. If proper crop-rotation was not followed, besides deterioration of soil fertility, pests and disease problems also come up.

8.10.3. Manure and fertilizer application

Balanced fertilizers and manure application based on soil test was an insurance against the drought. Better root-growth enabled plants to absorb moisture from deeper layers of soil. Singh and Ramakrishna (1974) reported that placement of nitrogen increased the yield of rainfed sorghum by 400 kg ha^{-1} and water use efficiency by 13 percent. Even in the year of aberrant weather conditions, balanced use of fertilizer and manure gave stable production. Placement of fertilizers particularly phosphatic and potassic ones, below the seed gave the best results.

8.10.4. Intercropping

Appropriate intercropping system e.g. soybean-pigeonpea, sorghum (hybrid)–pigeonpea, maize-soybean, maize-pigeonpea are more remunerative than mono-cropping even under adverse weather conditions.

8.11. SELECTION OF CROPS

Selection of crops for varying conditions of soil (texture, slope, depth, water holding capacity) and moisture availability periods have already been discussed. As regards selection of crops/varieties in view of aberrant weather, lot of valuable information on this was available from recent research conducted under All India Coordinated Dryland Research Project (ICAR) throughout the country including Indore and Rewa centres of Madhya Pradesh and other research projects of Agriculture Universities. As already stated, it had been established what was the best under particular weather conditions. But unfortunately the monsoon rainfall pattern could not be predicted precisely and as such it was not possible to plan cropping programme under rainfed conditions precisely in advance. The approach had to be to plan for normal season and then make mid-season corrections according to prevalent weather conditions. Also, the farmer should start with cropping practices having the least risk.

8.11.1. Rainfed crops for low moisture regime areas

Rainfed farming was once considered a very risky system of farming but modern dryland technology evolved by research had almost done away with the risk and recommendations were now available for the right crop/variety to be grown for successful rainfed farming. Sorghum, pearlmillet, pulses, groundnut and cotton had been found suitable for growing under rainfed conditions in low rainfall areas based on age-old experience of raising various crops by farmers. The suitability of these crops had been confirmed by modern dryland research conducted throughout the country. These crops adapted to moisture stress condition either due to their better root development enabling the crops to extract moisture from deeper soil layers or to their low transpiration rates or both.

8.11.2. Root depths of various crops

The data on active root-zone of different crops presented in table 47 indicated that onion, garlic and annual flowering plants were shallow rooted, millets, groundnut, maize, wheat and sorghum were medium rooted while cotton, pigeonpea, chickpea, linseed, soybean, pulses and safflower were deep rooted.

Table 47. Active root zone of different crops

Crop	Active root zone (cms)
Garlic	10-15
Onion	10-20
Millets	30-40
Groundnut	40-50
Maize	40-60
Wheat	50-70
Sorghum	60-75
Greengram/blackgram	60-80
Chickpea	100-150
Linseed	100-150
Mustard	120-150
Soybean	100-120
Cotton	150-200
Pigeonpea	150-200
Safflower	150-200

Pulses (blackgram and greengram) were medium deep rooted and thus could exploit soil moisture from deeper layers. However, because of profuse foliage, their evapo-transpiration was high and thus could not resist drought. Also, there were pests and disease problems not solved yet.

8.11.3. Moisture extraction pattern

Moisture extraction pattern of various crops from deeper soil layer (100-175 cms) was studied (table 48) and it was confirmed that chickpea and linseed extracted more moisture from deeper layers as compared to barley, wheat and oats (IARI, 1976). Bhargava et al (1976) studied soil moisture depletion pattern under maize and soybean crops and concluded that much more soil moisture was left in lower layers of black soil under maize crop.

Table 48. Moisture extraction by some crops from 100-175 cm layer

Crop and variety	Moisture extraction (percent)
Chickpea (NP 58)	26
Linseed (RR 9)	24
Barley (NP 13)	15
Oats (Kent)	13
Wheat (NP 718)	10

Source: A new technology for dryland farming IARI Bulletin 1976 p 121

8.11.4. Water use efficiency

Another criteria to judge the suitability of different rainfed crops was their water use efficiency. Water use efficiency (WUE) of a crop was its marketable crop produce per unit of water used as evapotranspiration (ET). That is, WUE = yield/ET. Water use efficiency of important *kharif* crops was studied and results were presented in table 49 (Dastane, 1968).

Table 49. Water use efficiency of different crops

Crop	Water used (ET) (mm)	Yield (kg ha^{-1})	Water use efficiency (kg mm^{-1})
Rice	1200	4500	3.7
Maize	675	4500	6.6
Pearl millet	500	4000	8.0
Sorghum	500	4500	9.0

It was apparent from the data presented in Table 36, that sorghum made the most efficient use of water, followed by pearlmillet and maize and that rice was the least efficient water user.

8.12. SELECTION OF CROP VARIETIES

Even within the same crop, there were some varieties/strains distinctly superior to others in their capacity to yield higher under limited moisture conditions. Such varietal capability was either due to their shorter growing period i.e. their adaptability to complete their life-cycle within comparatively shorter period or due to their rooting pattern enabling these to extract water for their growth from deeper layers of the soil or due to their ability to reduce evaporation/evapotranspiration by better canopy development and some stomatal mechanism or some deposit on leaves. One example of such a genetic wonder was creation of sorghum hybrids giving higher yields during shorter periods. Local varieties of sorghum required about 150 days to complete their life cycle while hybrid varieties only 80-90 days. Local varieties yielded about 15q ha^{-1} whereas hybrids 40-50 q ha^{-1}. Also, in drought years local varieties failed to flower and produce any grain whereas hybrids still yielded 40q ha^{-1} avoiding the drought during flowering and grain formation stages.

8.13. RAINFED CROPS FOR HIGH MOISTURE REGIME AREAS

Traditionally millets, pigeonpea, cotton etc. were largely grown in low rainfall areas, while rice and soybean in high rainfall areas. As already stated rice crop was the least efficient in water use. When grown on shallow and light soils, it suffered from drought even in high rainfall areas. As a matter of fact certain drought prone areas were identified with rice crop i.e. rice was not suitable crop for those areas. Soybean grows very well in high rainfall areas particularly on medium and deep black soils where waterlogging/water stagnation was not a problem or where both external and internal drainage were provided. Short duration varieties of soybean had proved very remunerative under rainfed conditions on black soils. These were adopted by farmers over large areas left fallow in *kharif* season previously making Madhya Pradesh a 'Soybean state'.

8.14. DATE OF SOWING

Sowing of *kharif* crops has to be decided considering the probable date of onset of monsoon in an area. Normally pre-monsoon showers were received before this date although their amount and distribution varied. Pre-monsoon showers were very useful for seed–bed preparation. Depending upon the quantity of pre-monsoon showers, surface soil got moist to certain depth.

8.14.1. Dry sowing

If pre-monsoon showers were less, the rain-water was used up in preparing the seed-bed. Consequently, the surface soil was dry and the question was whether to adopt dry sowing? Dry sowing was risky and the risk was greater when normal onset of the monsoon was late or when it was delayed. But early sowing of crops was profitable as yields were high. Also, on deep black soils with flat topography sowing of crops was delayed abnormally as after the onset of monsoon, the soil did not come in appropriate condition for sowing for long due to their high water holding capacity (WHC). Thus early sowing with least risk must be adopted. Farmers could start with dry sowing of early varieties of cotton on medium or deep black soil (with adequate surface drainage) or on alluvial soils. Dry sowing of crops like soybean and groundnut must be avoided as seed was costly and re-sowing might be unavoidable if rains after sowing were sufficient for germination but insufficient to sustain plant growth during the period before onset of monsoon. Even if the soil moisture was imbibed by the seed and it swelled, lack of moisture later on resulted in spoilage of seed and no germination took place. If pre-monsoon showers wetted the soil up to about 15 cms and normal break of monsoon was to take place after 6-8 days, taking risk of sowing was worth it. The depth of sowing must be adjusted so that seed was in moist soil. Even light showers after sowing were helpful to plant-growth, which would be harmful in case of dry sowing. Fertilizer dose based on soil-test must be placed below seed and usually it serves as an insurance against the drought.

8.14.2. Low plant population

Plant population obtained by dry/early sowing of *kharif* crops was usually not adequate, and there were gaps in between the plants. It was necessary to fill up these gaps timely to get high yield. The gaps could be filled using seeds of the same crop if the planting date was suitable for it. Sorghum was very sensitive regarding sowing date and its yield decreased considerably on late planting. The delayed planting attracted shoot-fly whose attack could destroy the crop completely. Therefore gap filling in sorghum if late should be done by seeds of another crop e.g. maize, lobia, pigeonpea, greengram, blackgram, sunflower etc. in that order. For cotton, seedlings can be prepared in polythene bags in advance and used for filling gaps when needed.

8.14.3. Sequence of sowing

Farmers should start with dry sowing of short duration variety of cotton on medium deep black soil or on alluvial soil and continue the sowing of sorghum and maize in moist soil layer created by pre-monsoon showers or by early onset of monsoon, while sunflower could be planted late in *kharif* (3rd or 4th week of August). This sequence of sowing various crops could also be practiced in case of aberrant weather due to which a crop could not be planted on its normal date of planting. Based on research work conducted under All India Coordinated Research Project on Dry-land Agriculture centres Indore and Rewa, following sequence of sowing dates of various crops/varieties was recommended to adopt when the onset of monsoon was late:

a. **When monsoon was late by two weeks:** Sorghum should not be sown when monsoon was late by two weeks. As an alternative, maize (varieties JM-812, Navjot and IVM-421) or soybean (varieties JS-93-05, Samrat, Monetta, NRC-7) should be planted on Vertisols and associated soils and greengram (varieties JM-721, Khargone-2, Pusa baisakhi) and blackgram (varieties JU-8-6, T-9) on light (mixed red and black and alluvial) soils.

b. **When monsoon was 2 to 4 weeks late:** Maize (varieties already stated) on light soil, pigeonpea (varieties Jagrati, Pragati, Asha) on deep soil, and sesame (varieties JT-21, JT-55 and JTS-8) should be planted in that order.

c. **When monsoon was greater than 4 to 6 weeks late:** Sesame (varieties already stated) and Sunflower (varieties already stated) should be sown.

d. **When monsoon was greater than 6 to 8 weeks late:** Safflower (var. Morden, Surya, Manjira), cowpea (varieties Gomti, Pusa Komal, Pusa dofasali), castor (varieties Gauch, Varuna) should be planted.

e. **When monsoon was greater than 8 weeks late:** Toria (variety JT-1), Castor (variety Varuna) on alluvial soil and safflower [varieties JSF-1, JSF-7(spineless), JSF- 73, Sharda] on deep black soil, must be sown.

8.14.4. Effect of sowing date on maize yield

As already stated, dry sowing of sorghum and maize (before onset of monsoon) was risky but highly profitable. Results of experiments conducted at maize research station, Chhindwara, Madhya Pradesh during 1964, 1965 and 1967 revealed that sowing maize 20 days earlier than normal date (3rd week of June) gave the highest yield (78.9q ha^{-1}) while sowing delayed by 20 days resulted in lowest yield of 47.9 q ha^{-1} (Table 50).

Table 50. Effect of date of planting on maize yield

Date of planting	Yield (q ha^{-1})
20 days before normal date	78.9
10 days before normal date	75.5
Normal sowing (3rd week of June)	61.4
10 days after normal date	58.1
20 days after normal date	47.9

8.14.5. Effect of date of planting on sorghum yield

Experiments conducted under all India Coordinated Sorghum Improvement Project, Indore Madhya Pradesh during 1970-72, showed that sowing sorghum just one week earlier than normal planting date (with onset of monsoon) increased the yield of sorghum by about 25 percent while sowing delayed by two weeks reduced the yield by about 60 percent (Table 51).

Table 51. Effect of date of sowing on sorghum yields

Date of sowing	Yield (q ha^{-1})
One week before normal date	40.4
Normal date	32.7
One week after normal date	28.1
Two weeks after normal date	12.6
Three weeks after normal date	9.1

Thus, early sowing was more profitable in case of sorghum. Also the yield was reduced miserably if sowing was delayed by two weeks, particularly because of serious shoot-fly attack in late sown sorghum crop.

8.15. DROUGHT PRONE AREAS

An area where occurrence of drought was frequent was called drought prone area i.e. an area where 2 to 4 years out of 5 years were drought years. Survey of drought prone area conducted by different organizations/agencies has reported different figures of drought prone area in India. The Irrigation Commission (1972) identified 326 talukas (a revenue unit) as drought prone areas covering 49.7 million ha in 67 districts. The National Commission on Agriculture (1976) and Central Water Commission (1978) adopted a slightly different criteria and termed only 315 talukas as drought prone covering an area of 51.13 million ha in 74 districts (Singh and Samra 2006). No drought prone areas were reported in Bihar, Jharkhand and Chhattisgarh states. Eastern Madhya Pradesh and eastern Uttar Pradesh fell under semi-arid (moist) and sub-humid (dry) regions and Chhattisgarh state under sub-humid (moist) and humid regions of the country. Also the probability of drought was low-once in five years or more. The drought prone areas in these states were shown in Figure 20. In all, 26 talukas (spread over 3.73M ha) in Madhya Pradesh and four talukas (spread over 0.46M ha) in Uttar Pradesh were declared as drought prone area (Singh and Samra, 2006).

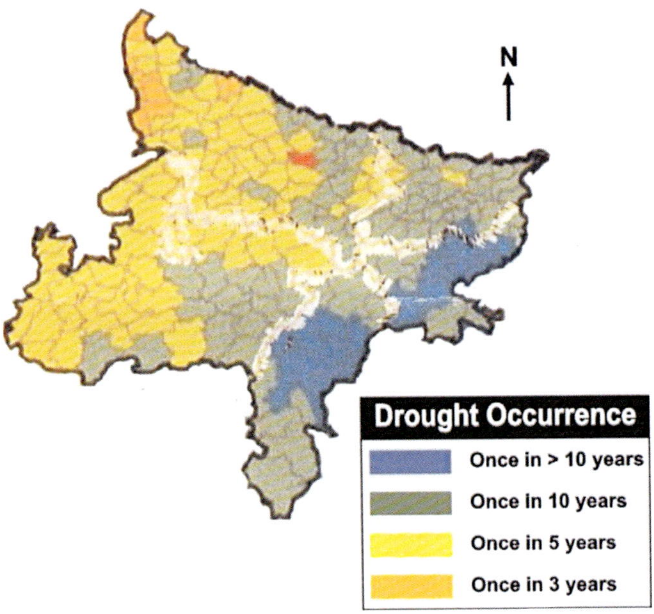

Fig. 20. Drought prone areas

8.16. WHAT IS DROUGHT?

In dictionaries, the term drought was defined as a long period without rain during the plant growth period, or as an extended period of dry weather injurious to crops. Drought was the condition under which not only plant and animal life but human beings also suffered from lack of adequate water–supply.

That is water was not available whenever needed upto requirement. Drought is a meteorological term involving a rainfall deficit for a period of time long enough to cause moisture depletion in soil and water deficit with a decrease of water potential in plant tissues. According to Sinha (1986) agricultural drought was the inadequacy of water availability including precipitation and soil moisture storage capacity, in quantity and distribution during the life cycle of the crop plant which restricts the full genetic potential of the plant. For farmers drought means low rainfall or a dry spell causing low crop-yields or crop failure depending upon the severity of drought. American Metrological Society defined drought as a period of abnormally dry weather sufficiently prolonged to cause a severe hydrological imbalance in the area affected. Drought differs from aridity because aridity was normal phenomenon of low rainfall areas whereas drought was abnormal phenomenon resulting from aberrant weather and could occur in low rainfall as well as in high rainfall regions.

8.16.1. Classification of droughts

Droughts were broadly classified into 3 categories, viz. Meteorological, Hydrological and Agricultural.

a. **Meteorological drought:** Meteorological drought was known as a situation in which there was a significant decrease (>25 percent) of normal rainfall over an area. Such drought, if prolonged, led to marked depletion of surface and ground water.

b. **Hydrological drought:** Meteorological drought if prolonged was hydrological drought and this resulted in marked depletion of surface and ground water, i.e. drying of tanks, reservoirs, etc. took place.

c. **Agricultural drought:** When both rainfall and soil moisture together could not meet the requirement of a healthy crop during crop growing season, such drought was termed as agricultural drought.

8.16.2. Drought *vs.* total rainfall

Drought was not always related with total rainfall. Since drought was related with moisture availability, it might not always be related with total rainfall. The rainfall distribution pattern was more important than the total rainfall. The total rainfall received at a place may be low but if distributed well, crop plants may not suffer from moisture deficiency and consequently no drought that is if showers were received whenever moisture deficiency was felt, the crop-growth did not suffer. Also besides the rainfall, nature of the crop and type of soil (its moisture storage capacity) were important factors. The water requirement of crops varies and accordingly at a particular place and with the same rainfall distribution, rice might suffer from drought but not sorghum as water requirement of rice was much higher than that of sorghum. Similarly, on a light and shallow soil sorghum crop could mature in any year but not the rice crop. Further, the average total annual rainfall in undivided Madhya Pradesh varied from 700 mm in west to 1600 mm in east and climate was categorized semi-arid to sub-humid. Still droughts were there with varying frequency in different areas.

8.16.3. Frequency of drought

During the period 1885-2005, India experienced 22 major droughts, of which the severest five occurred in 1899, 1918, 1972, 1987 and 2002. During 2002, 22 million ha area could not be sown, 47 million ha of the sown area was damaged and food production got reduced by 29 million tonnes. There was widespread damage to coconut, arecnut, banana, grape, pomegranate, mango, citrus crops etc. even in high rainfall coastal areas. Drought could occur in high rainfall areas also because agro-ecological systems in such areas were in equilibrium with high moisture conditions and any deviation from normal rainfall could lead to drought in the areas. The probability of drought in different states estimated on the basis of long term rainfall data was presented in Table 52.

Table 52. Probability of occurrence of drought in different states/regions

States/Regions	Frequency of 75% or less rainfall
Assam	Rare, once in 15years
West Bengal, Chhattisgarh, Konkan, Bihar, Jharkhand, eastern Uttar Pradesh and Orissa	Once in 10 years
South interior Karnataka and Vidarbha	Once in 5 years
Gujarat, eastern Rajasthan, south western Uttar Pradesh and south western Madhya Pradesh	Once in 3years
Western Rajasthan	Twice in 5 years

Thus, even in humid areas drought could occur although the probability of its occurrence was low as compared to sub-humid and arid areas. Further, the frequency of drought in central India was very low except Jhabua hills agro-climatic zone in Madhya Pradesh and Bundelkhand region of Uttar Pradesh.

8.16.4. Intensity of drought

Not only the frequency but intensity of drought also varied in different years and at different places. Drought might be mild, moderate or severe depending upon the damage done by the drought. The intensity of drought could be judged by moisture deficit as it affected the plant growth. Intensity of drought was a ratio of actual evapo-transpiration to potential evapo-transpiration (AET/PET) during the crop-growth period. Based on this ratio, the intensity of drought was shown in Table 53.

Table 53. Intensity of drought based on AET/PET ratio

AET/PET Ratio	Severity of drought
< 0.24	Severe
0.25-0.49	Moderate
0.50-0.74	Mild
> 0.75	Low

8.16.5. Sample survey

Sample survey was the scientific basis for determining the extent (area) and intensity (degree) of the drought. The role of field workers was very important in correct assessment of severity of drought based on the extent of damage done to farmers' crops by the drought e.g. 25 percent, 50 percent, 75 percent or 100 percent (complete failure of crops). Also, once the damage has been done, not much can be done except relief measures.

8.17. DROUGHT MANAGEMENT

As already stated, once the damage to crops by drought has been done, farmers themselves are forced to adopt economic measures. They are used to reduce consumption, postpone social engagements such as marriages and other functions. If possible, they migrated to other areas where situation was better weather-wise, with their livestock, or they sold their livestock. They also took consumption loans and sold their assets like gold and silver ornaments to cope with drought. Only very few farmers could stock food grains and fodder produced in favourable years for use during drought years. However, based on research work conducted under Dryland Research Project and various Crop Improvement projects, certain measures could be suggested to minimize or even eliminate the damage to crops in drought prone areas. For this, advance precautionary measures, mid-season correction measures and long term measures have to be taken.

8.17.1. Advance precautionary measure

Since in drought prone areas any year could be a drought year, certain precautionary measures had to be taken in advance e.g. seed treatment, crop rotation, fertilizer and manure application, intercropping, selection of crops/varieties, early but safe date of sowing, gap filling, etc. as discussed already.

8.17.2. Mid-season correction

If the drought stroke when the crop was standing in the field, mid-season corrections were necessary to save the crop. For this, life-saving irrigation could be applied, if available. If no irrigation was available, an effort had to be made to reduce the soil moisture loss to the minimum possible through following measures:

8.17.2.1. Use of anti-transpirent: Anti-transpirents close the stomatal openings of the leaves and thus reduce transpiration, e.g. hydroxylamine hydrochloride phenyl mercuric acetate, silicone and acetyl and stearyl alcohols.

8.17.2.2. Use of light reflectants: Some substances when sprayed on plants to form a thin film over plant leaves reflect the light and reduce the temperature of the plant and consequently the transpiration e.g. spray of kaolinite and calcium chloride. Studies to mitigate early and late season droughts were conducted under All India Coordinated Research Project on Dryland

Agriculture at Agra, Dantiwada and Bangalore centres. Spray of 5 percent kaolin on pearlmillet proved beneficial at Agra centre and on fingermillet at Bangalore centre. Also spray of 2 percent solution of $CaCl_2$ on pearlmillet at Bangalore and five percent solution of $CaCl_2$ on pearlmillet at Dantiwada was found promising (Singh and Samra, 2006).

8.17.2.3. Use of plant-modifiers: Certain materials when sprayed on plants were reported to modify their structure so as to lose much less water from their surfaces e.g. Cycocel (2 chloroethyl trichloromethyl ammonium chloride) or CCC, spray dwarfened the plants and thus reduced the total surface area and thereby water-loss. Cycocel spray was found very effective when sprayed on cotton- one to three times at intervals of 15 days each if necessary. In Australia, the application of CCC on wheat prior to grain setting increased the number of grains per ear head and thereby the yield (Arnon 1975). In Israel, plants treated with CCC remained turgid and survived longer than control under moisture stress conditions. When the soil moisture reached wilting point stage, growth of treated plants exceeded that of untreated ones (Halvey and Kessler 1963). In India also, foliar application of CCC increased yields of wheat significantly (Shrivastava et al 1968).

8.17.2.4. Use of anti-evaporants on soil: A large quantity of moisture was lost from soil by evaporation, specially from bare moist soil i.e. soil not covered by crop-canopy. A number of substances when added to soil were reported to reduce evaporation from the soil, for example, plastic emulsions, Encap, Suncap, Jalshakti, etc.

8.17.2.5. Use of mulches (mulching): Different type of mulches reduced loss of moisture from soil. Mulching practices varied from physical manipulation of surface soil to application of organic materials including plastic sheet as a cover on inter-row space. Various types of mulches are discussed in continuation.

8.17.2.5.1. Dust Mulches: Repeated intercultural operations in the standing crops were recommended for destroying weeds. Besides killing the weeds, such operation stirred the surface and created layer of loose and pulverized soil on the surface which prevented loss of soil moisture from lower horizons as capillaries through which moisture came up and got lost to atmosphere were disturbed. Thus this loose soil served as dust mulch. The practice of dust mulch was more effective on Vertisol where cracks were formed on drying of the soil and got wider and deeper with time with continuous loss of soil moisture from deeper layers. This loss of moisture from lower layers was almost prevented as formation of cracks was nearly checked by "dust mulch".

8.17.2.5.2 Plastic mulch: Spreading a plastic sheet on inter-row space prevented loss of soil moisture to atmosphere as it was impervious to vapour movement. The soil moisture rising to surface through capillaries got condensed at surface due to lowering of temperature of plastic sheet at night. Thus, it increased the amount of moisture available to plant, which was the objective of mulching.

8.17.2.5.3 Organic mulch: Spreading of organic material on inter-row space checked evaporation and thus served as mulch. Depending upon their availability locally, various materials such as wheat and rice straw, stalks of maize, sorghum and pearlmillet, stover of soybean, blackgram, greengram, etc. crop residues such as stubbles of sorghum and maize, stalks of mustard, safflower, sunflower and pigeonpea, and waste materials such as trash, dry grass and weeds, and material available on threshing of maize cobs and ear heads of pearlmillet and sorghum, can be used for mulching by spreading the material on inter-row space. The layer of mulch on surface soil prevents sun-rays from hitting it directly and thereby minimizes the evaporation loss of profile stored moisture. Studies on organic mulches conducted at Central Arid Zone Research Institute, Jodhpur, Rajasthan showed that on application of organic mulch, the soil temperature at 10cm depth was reduced upto 6°C and moisture content increased. Similarly, dry grass mulches at the rate of six t ha^{-1} reduced the maximum soil temperature, increased soil moisture content and resulted in 40 percent increase in yield of pearlmillet (Gupta, 1980).

8.17.2.6. Comparison of different types of mulches: Studies were conducted at All India Coordinated Research Project centre, Indore, to compare different types of mulches used in soybean – safflower crop sequence and the results are presented in Table 54.

Table 54. Effect of different mulches on soybean and safflower yield (kg ha^{-1})

Treatment	Yield (kg ha^{-1}) Average of 3 years (2003-2005)			
	Soybean	Increase Over no mulch (percent)	Safflower	Increase Over no mulch (percent)
No mulch	1605	-	594	-
Dust mulch	1779	11	905	52
Polythene mulch	1990	24	1142	92
Straw mulch (2t ha^{-1})	1905	19	1248	110
Weed biomass (5t ha^{-1})	1972	23	1097	85
Glyricidia (2t ha^{-1})	1934	20	1117	88

Source: Jain et al (2014)

The data presented in table 54, bring out clearly the useful effect of various types of mulches on both soybean and safflower yield and that effect on safflower yield was higher than on soybean yield. Apparently soil moisture conservation by use of mulch was more important in *rabi* season than in *kharif.* Also, the effectiveness of mulches in case of safflower was observed right from seeding time when moisture content of soil was higher in mulched plots as compared to no mulch plot, and this resulted in better germination of safflower seeds and consequent better plant population in mulched plots. Percent increase in yield of safflower was also higher (52 to 110 percent) than that of soybean (11 to 24 percent). This showed that soil moisture regime in *rabi* was below optimum for safflower growth and any improvement in it through mulch resulted in much increased yield, whereas without mulch the moisture stress decreased the yield as compared to what it would have been in case of favorable moisture regime throughout the growth period. In rainy season the moisture regime of soil was much better and it became unsatisfactory only during long dry spells between the rains particularly because of uneven distribution of rainfall. Use of mulches in soybean crop reduced the adverse effect of uneven distribution and increased soybean yield. Also, if during any year the distribution of rainfall was uniform throughout the growing period of soybean crop, no significant difference in soybean yield was observed due to mulching/different kinds of mulches.

8.17.3. Reducing plant population

As the amount of moisture stored in soil was limited and there was no chance of its recharge due to drought, reducing plant population (thinning) helped the remaining plants to attain maturity. The extent of thinning depended on the degree of drought. Alternate plants in the row or alternate rows could be removed to reduce the plant population to 50 percent. Removing rows and stirring the surface soil with a hand-hoe created a layer of loose soil on the inter-row space and served as dust mulch to conserve the soil moisture.

8.17.4. Ratooning of some *kharif* crops

When drought is so severe as to kill or almost kill the crops, some crops can be ratooned e.g. in case of sorghum, vegetative part can be harvested leaving stubbles in the fields for regeneration with late rainfall.

8.17.5. Planting other crop on failure of originally sown *kharif* crop

The crop will be chosen according to the time when sowing can be done e.g. sunflower in August or later on (*i.e.* late *kharif*), toria or safflower in September, and linseed and chickpea in October (2nd/3rd week). Also, to face the drought better it was necessary to plant perennials on field bunds as perennials, in general, were comparatively more resistant to drought than annuals.

8.17.6. Defoliation of crop-plants

To reduce water-loss by transpiration of crop-plants, their defoliation can be practiced in case of severe drought. However, it has not been found useful in every situation or always, because defoliation, actually decreased crop-yield potential after the revival of rains in the season.

8.18. LONG TERM MEASURES OR PREVENTIVE STEPS

Droughts are man-made usually and at least in central India. The rice crop grown on eroded/degraded soil with low water holding capacity and depleted of nutrients, failed frequently and blamed the drought. Some long term measures or preventive steps against the drought were:

i. Management of rain-water on water–shed basis, including run-off collection and recycling, as discussed in previous chapters.

ii. Appropriate crop-plans according to suitability of land/soil.

   a. Growing crops/varieties to suit the moisture availability period

   b. Change in cropping strategy in view of prevalent weather conditions.

   c. Alternate land – use if there were frequent droughts.

■■■■

IX

FARM MACHINERY AND IMPLEMENTS

In dryland farming, it was necessary to complete all the activities from field preparation to post-harvest operations after the harvest of crops, in the least possible time. It is possible only through the use of appropriate equipment and tools which do away with the drudgery of work and increase the efficiency of farm-labour rather than replace it. Hence, the importance of development of improved bullock-drawn implements, farm equipment and tools cannot be over emphasized.

The urgency of the development of farm machinery and implements appropriate to the needs of dryland farmers has been taken in right earnest and a Central Institute of Agricultural Engineering has been established by the Indian Council of Agricultural Research (I.C.A.R.) at Bhopal, Madhya Pradesh. Fortunately, some work on bullock drawn implements and hand-tools had been undertaken under the All-India Coordinated Dryland Research Project centres Indore and Rewa also. As a result of this, some farm equipment and tools were found useful to dryland farming in central India and the same are detailed as follows.

9.1. TILLAGE

Tillage was defined as mechanical manipulation of soil to improve its physical condition for better crop-production, i.e. to provide soil environment favorable for germination of seed and for growth and development of plants resulting in higher yield. According to Gajri et al (2012) tillage was a process in which human, animal and/or machine energy was applied for physical manipulation of soil to provide soil environment favorable for the growth and production of plants. Again, production of plants involved a sequence of events including seeding, seed-germination, seedling emergence and development to the stage of maturity. Traditionally, the tillage process involves a series of primary operations to break the soil surface into clods and also a series of secondary operations to pulverize the soil into good tilth. Tillage requirement was highly soil and crop specific and basic objective of tillage was to be obtain better soil environment (soil tilth) for crop production. Tillage of the soil under dryland/rainfed farming was important for (i) weed control, (ii) infiltration of water into soil particularly in medium and deep black soil, (iii) obtaining good seed-bed having suitable soil tilth for different crops for germination of seed, and (iv) conservation of soil moisture by creating a layer of loose soil on the surface.

Recently doubts have been raised as to whether conventional method of tillage for various crops and for wheat in particular, results in superior crop yields. Research had shown this might not be true. Preparing a fine seed-bed before sowing of wheat may not be necessary and could be a waste of time and energy. Repeated ploughing and planking to obtain a very fine and fluffy seed-bed may not always lead to better germination and crop-stand and consequent higher yield.

9.1.1. Preparation of good seed-bed

A good seedbed had optimum tilth, was weed-free and had no mechanical impedance to root growth. The optimum tilth provided desirable air-water relations in soil for germination of seeds and favorable environment for growth of root and shoot i.e. there was minimum resistance to penetration of root, and air-water relations were optimum for development of plant after germination of the seed. Depending upon the kind and number of tillage operations undertaken, the surface soil was cut into pieces of different sizes (clods), loosened, stirred and pulverized. Besides kind and number of tillage operation, the resultant tilth of the soil depended upon the soil moisture content at the time of tilling. Also, the moisture content was more critical in case of black soils (clay to clay loam texture) than alluvial soils. Black soils (Vertisol and associated soils) if tilled when wet, got puddled i.e. they lost their structure, got compacted and set hard on drying. On the other hand, if tilled when too dry, the energy required was much more and hard clods were formed on tilling. The optimum moisture content for tillage was when the soil was in a friable condition. Under rainfed condition, fineness of appropriate seed-bed varied according to the crop to be sown/planted. For rainfed wheat and mustard crop, very fine seed-bed was prepared when those crops were grown after *kharif* fallow.

9.1.2. Weed control

An important objective of tilling the soil was to obtain weed-free seed-bed. Ploughing, cross ploughing, bakharing, etc. after rains uprooted the germinated/ regenerated weeds, incorporated these into soil to decompose and thereby improve the soil-structure. It also mixed the Farm Yard Manure/compost placed on the field in small piles, and stirred the soil, all for good tilth. It was mainly because of weed-control function that the conventional tillage was still popular with farmers.

9.1.3. Improvement of soil properties

Some soils developed plough sole as stirring of soil by tilling was restricted upto a particular depth. Also, some soils had a problem of crusting (formation of an impermeable crust on surface of soil). These problems were overcome by tillage operations.

9.2. VARIOUS TILLAGE SYSTEMS

Depending upon the intensity of tillage operations undertaken for establishing a crop, tillage systems are classed into (1) No tillage (2) Reduced or minimum tillage (3) Conservation tillage and (4) Conventional tillage.

9.2.1. No tillage

No tillage system or zero tillage system meant planting crops without preparing a seed-bed. The seed was put directly into bare soil or under crop-residues using a suitable equipment. The crop-residues are retained on soil-surface i.e. these are not removed or burnt, and seed was drilled below crop residues. The seed was broadcast also in a field in standing rice crop or immediately after harvesting. There was a common practice of *utera* in Chhattisgarh state where lathyrus or chickpea seed was broadcasted in rice fields to grow a crop in *rabi*. This practice was also being followed in the rice growing areas of Nepal where wheat was grown after paddy by broadcasting wheat seed in wet/moist soil with stubbles of recently harvested rice crop. In the northwestern part of India, planting of linseed immediately after rice-harvest in untilled wet soil was a common practice. The weeds are controlled by using herbicides or manually (Gajri et al 2012). There was lot of emphasis on no-tillage or minimum tillage these days particularly for rice-wheat crop sequence advocated after extensive research effort by Rice-wheat consortium for Indo-Gangetic Plains (RWC-CIMMYT 2003). Repeated traditional tillage operations escalate the cost of cultivation of wheat and reduce the net profit availed by farmers (Harrington et al 1993). It was generally believed that if tillage operations prior to wheat sowing were not performed, crop-yield would suffer because of poor crop-stand and also due to poor soil aeration following an irrigation or rainfall. But the research has shown that zero-tillage establishment of wheat after rice gave as much or higher yield of wheat as compared to normal tillage sown crop. This was explained mainly by timely planting in zero tillage plots. However, the results are specific for Rice-wheat crop-sequence possible only under irrigated conditions. Under rainfed condition, rice was not an appropriate crop for light and shallow soils even in high rainfall areas. Also, fallow-wheat system was the least remunerative crop-sequence under dry farming conditions as already discussed. However, zero tillage/no tillage was important in rainfed farming also when next crop in *rabi* season was to be sown as early as possible to avoid moisture loss through tillage or evaporation from moist soil, e.g. under sorghum-chickpea, maize-chickpea, soybean-chickpea crop-sequences under rainfed farming. Under these conditions, sooner the sowing was done, greater was the utilization of soil-profile-stored moisture and consequently the yield.

9.2.2. Reduced or minimum tillage

Reduced or minimum tillage system referred to a system in which one or more tillage operations (necessary for removing crop residues) were eliminated from conventional tillage system. That is, only those operations were undertaken which were absolutely essential for crop production under specific soil and crop conditions. When these operations were reduced to minimum for establishing a crop, it was minimum tillage system. Thus, reduced tillage system was less intensive than conventional tillage system where as the minimum tillage system was the least intensive and could be close to 'no tillage system' under certain circumstances, for example, tilling and planting in just one operation. In fact, several variations of reduced tillage system-almost no tillage to nearly conventional tillage, are in vogue globally (Unger 1984). The advantages of reduced/minimum tillage over conventional tillage were: (i) saving of time and energy and conversantly the cost of cultivation, (ii) conservation of soil and water. However, there might be problems of diseases and pests from crop-residues and improper seed-placement through crop-residues.

9.2.3. Conservation tillage

Conservation tillage was defined as the system of tilling in which crop residues-stubbles, roots and leaves, were retained on soil surface and tillage operations were not so intensive as under conventional tillage. Thus, it was reduced tillage or minimum tillage which did not destroy or burry plant residues, i.e. the tilling operations were conducted only on inter-row space of the harvested crop, without disturbing the crop residues. Also, tilling and planting in a single operation without disturbing the crop residues was conservation tillage. The main advantages of conservation tillage were: (i) build-up of organic matter content of the soil, and (ii) reduction in cost of cultivation of crops (Tomar 2012). But there were some disadvantages also, e.g. the crop-residue provided a refuge to pests and pathogens and could invite termites.

9.2.4. Conventional tillage

The field preparation started right from summer season when deep ploughing was undertaken, particularly if perennial and obnoxious weeds such as kans (*Saccharum spontaneum*), jharberi (*Zizyphus jujuba*), baisurai (*Pluchia lanceolata*) etc. were infesting the fields. After summer cultivation, tillage operations were again started during pre-monsoon rains. Ploughing, cross-ploughing on alluvial soils and bakharing, cross-bakharing on black soil were undertaken using bullock power. Where bullock power was replaced by tractors, tractor driven cultivators/harrows were used. This preparatory tillage killed weeds germinated by pre-monsoon showers and created a layer of loose soil on field-surface. The depth of loose soil-layer varied from 5 to 15cms depending upon the tillage equipment used and soil type. For all *kharif* crops except paddy, the seed-bed so prepared was satisfactory. For paddy, the field preparation varied according to the methods of planting, viz. (i) broadcasting seed and *biasi* operation after seedlings emerged, (ii) drilling seeds in rows, and (iii) transplanting seedling raised in nursery. Following conventional tillage for *rabi* crops, the farmers continued ploughing and cross-ploughing, bakharing and cross bakharing throughout the monsoon season and ploughing/bakharing and planking after the monsoon was over. These operations were repeated till a fine seed bed was ready for sowing wheat and mustard at an appropriate date. For other *rabi* crops (rainfed), comparatively coarse seed-bed was alright. For gram crop, even a seed-bed with small clods was satisfactory. It was now recognized that the conventional tillage operation adopted by farmers for getting fine seed-bed for wheat and mustard were excessive and not necessary.

As pointed out already, tillage practices by farmers vary widely. In certain cases, conventional tillage practiced by farmers may be more intensive than required, e.g. that for preparation of seed-bed for wheat and mustard, using traditional bullock power, when these crops are raised after *kharif* fallow. But reducing tillage for controlling the weeds by using weedicides may not get popular with the farmers, particularly the resource-poor ones and due to loud voices against the use of chemicals by advocates of organic farming and by environmentalists. Further, zero-tillage recommended for growing wheat after paddy, was not practicable under rainfed/dry farming conditions due to high water requirement of wheat and comparatively low capability of wheat plant to utilize soil moisture in lower horizons due to its fibrous root system. That is why farmers continue to practice conventional tillage.

9.3. CONVENTIONAL TILLAGE VS. MINIMUM TILLAGE

The comparative effect of these systems on soil erosion and soil productivity was discussed in continuation.

9.3.1. Soil Erosion

The conventional tillage consisting of primary and secondary tillage operations normally performed, has been blamed of being expensive, causing higher run-off and consequent greater soil and nutrient losses under maize-toria cropping system (Singh and Sharma 2008). As an alternative technology to conventional tillage, conservation tillage or minimum tillage and even no tillage were being emphasized for specific cropping systems. Conservation tillage refers to minimum soil manipulation necessary for crop-production under specific soil and climatic conditions. While conventional tillage aimed at a clean seed-bed removing crop-stubbles etc. completely (up-rooting crop-residues and burning these), conservation tillage maintained crop-residues in situ, as a cover on surface soil to protect it against erosion and build up its fertility and organic matter status which improved soil structure (Table 55). Further conservation tillage technology was less expensive compared to conventional tillage.

Table 55. Effect of tillage on run-off and soil loss

Type of tillage	Run-off (%) of rainfall)	Soil loss (tonnes ha^{-1} $year^{-1}$)	Nutrient losses (kg $ha^{-1}$$year^{-1}$)		
			N	P	K
Conventional	23.3	17.95	32.1	0.57	13.3
Minimum (+crop residues)	20.6	13.58	18.5	0.41	8.8

Thus, minimum tillage leaving crop-residues on soil surface reduced run-off, soil-loss and loss of plant nutrients from soil.

9.3.2. Effect on soil-productivity

In a study by Central Soil Water Conservation Research and Training Institute, Dehradun, minimum tillage along with incorporation of crop residues resulted in higher yield of maize and toria, by 8 percent and 9 percent, respectively, as compared to conventional tillage. The treatment increased contents of organic carbon, total nitrogen, available P and K by 21, 16, 22 and 8 percent, respectively. It also improved bulk density of soil from 1.36 to 1.30 g cm^{-3}, and infiltration rate from 6.4 to 7.2 mm hr^{-1} (Table 56).

Table 56. Effect of minimum tillage + crop residues on soil properties

Treatment	pH	Organic carbon (%)	Total N (%)	Available P (kg ha^{-1})	Available K (kg ha^{-1})	B.D. (g cm^{-3})	Infiltration rate (mm hr^{-1})
Untreated (Initial status)	5.1	0.58	0.068	32	136	1.36	6.4
Conventional tillage	5.4	0.63	0.070	32	136	1.36	6.8
Minimum tillage + crop residues	5.2	0.70	0.073	39	147	1.30	7.2

9.3.3 Effect on organic matter distribution

Repeated tillage operations executed under conventional tillage system increased soil aeration and the consequent microbial decomposition of soil organic matter. Hence, in the long run, the content of soil organic matter was reduced by conventional tillage. Blevins (1986) observed a marked decrease in organic matter content under conventional tillage system as compared to that under 'no tillage' (Table 57).

Table 57. Effect of tillage systems on soil organic matter content (percent)

Soil depth (cm)	Soil Organic Matter Content (percent)	
	No tillage	Conventional tillage
0-5	4.82	2.40
5-15	2.34	2.31
15-30	1.15	1.22

9.4. TILLAGE IMPLEMENTS

Bakhar (Blade Harrow) for black soils and deshi plough for alluvial and other soil types were the most commonly used bullock drawn tillage implements. Tractor drawn implements were also used by many farmers thanks to custom-hire service now getting popular in villages e.g. Mould board plough, Disc plough, cultivators and harrows. These implements were used with specific purpose and under specific conditions. Tractor driven mould-board plough was used for deep ploughing to eradicate perennial obnoxious weeds such as kans (*Sacharum spontaneum*), Jharberi (*Ziziphus nummularia*) etc. Duck foot cultivator was found very suitable for black soils at Indore and could be adopted over larger areas in Madhya Pradesh and in Central India.

9.4.1. Bullock/Animal Drawn

9.4.1.1. ***Deshi plough*****:** Deshi plough was an ancient but still the most commonly used bullock drawn tillage implement, which had stood the test of time, without much modification by the scientist. The main body of the plough was a wooden block having broad lower part with two holes (over one another) and tapering upper part with a hole for fixing the handle. An angular wooden block with an iron share was fixed on the lower most hole of the plough body while a long wooden beam was fixed on the other (upper) hole. The beam had three holes for hinging the plough to a wooden yoke made to fit on the neck of two bullocks/he-buffaloes/other animals. The three holes on the beam (upper part) provided for depth adjustment of the plough. The plough was in use on all soil-types except Vertisol and associated soils. The deshi plough was not very good and efficient tillage implement as it opened only V-shaped furrow and normally left an inverted V-shaped (Λ-shaped) unploughed land. Also it did not invert and mix the soil, and therefore it was less effective in destroying the weeds. However, on ploughing and cross-ploughing, Λ-shaped unploughed land was taken care of. Further, its versatility to work as a harrow, as a puddler, as a cultivator, as a sowing device (with a funnel type seed dropper) and as a digger to harvest root crops like potato, sweet potato, etc. kept the deshi plough still popular with the farmers, particularly the small ones. These ploughs were normally manufactured by village carpenters/black smith. These varied slightly in their designs from state to state, although the basic components remained the same (Fig. 21, 22).

Fig. 21. Deshi Plough

Fig. 22. Deshi Iron Plough

9.4.1.2. Mouldboard plough: Mould board plough – an animal drawn and walking type, opens up and turns under the surface soil i.e. it inverts the soil and was therefore suitable for incorporating the crop residues into soil. But in black soil its power requirement was too much to be suitable for animal to drive (draw) it. Also, it had problems in hard stony soils and soils having underground obstructions. When used on alluvial, red, red and yellow soil, back furrows (ridge) and dead furrows (ditch) are formed, and thereby level of the field was disturbed. Animal drawn mouldboard plough was normally single bottom type 15 to 25 cms size (Deshpande, 1973). Various makes such as Kirloskar No. 11, Meston and Shabash were available and these were found useful for eradicating perennial weeds such as doob (*Cynodon dactylon*) and jharberi (*Zizyphus jujuba*), etc.

9.4.1.3. Improved Bakhar (Blade harrow): The bakhar was also the most commonly used animal drawn tillage implement for shallow cultivation of black soils (Vertisols and associated soils). This traditional implement consisted of a 1.5m long rectangular wooden block to which beams (wooden) and bakhar-blade (made of iron) were fixed. Two wooden beams were fixed to the wooden block such that these converged to meet together. The beams were hinged to yoke as in case of plough, for using the bakhar. The bakhar has been improved by modifying the iron blade which was with a straight cutting edge and consequently needed greater force to operate. This traditional blade was replaced by V-blade and an additional cylindrical part was added to work behind the blade to break clods and level the field. The V-blade cut the soil with comparatively less force and the resulting seed-bed was finer and more even, as compared to that obtained with traditional bakhar. The improved bakhar had effective breadth (of tilled soil) 50cms, effective depth (of pulverized soil) 6.0cms, and working capacity of 0.062 to 0.075 ha hr^{-1}. The estimated cost of the implement was Rs. 4000 approximately (Dubey and Singh 2014).

9.4.1.4. Patela or patta: It consists of a long (2.5m) rectangular and heavy solid wooden block with two hooks for tying ropes to connect the patta/patela to yoke for dragging it. It is used on ploughed field to break clods and level the field. In post-monsoon operations to prepare seed-bed for *rabi* crops, levelling by pattaing conserves soil-moisture.

9.4.1.5. Patela-harrow: This implement was useful for breaking soil-clods and collecting crop-residues (stubbles, etc.) when used after ploughing the field. It consists of a wooden plank (teak or good-wood board), hooks for collecting crop-residues, a lever and a wheel with notches as a device to raise hooks and fix them in a particular position for effective working. The size of wooden plank was 1.5 to 2.0 meter in length and 50 to 60 cms in width. The effective depth of its operation was 6 cms, and its working capacity was 0.30ha/hour. Its estimated cost was Rs. 6000 (Dubey and Singh 2014).

9.4.1.6. Puddling machine: Puddling of soil before transplanting paddy seedlings was a recognized practice for wetland paddy cultivation. A puddler or puddling machine was essentially a special kind of harrow which churned the soil in standing water and buried the standing weeds and crop-residues, if any. Puddling dispersed the soil particles in standing water, made the soil impermeable and compact by reducing its apparent specific volume and thus facilitated transplanting. Conventionally farmers used a deshi plough or *datari* or *khutia* for puddling operation. The improved puddlers, both animal and tractor drawn, were now being manufactured indigenously. These puddlers worked on the principle of high mechanical dispersion. Both animal and tractor drawn cage wheel puddler consisted of main body, driving mechanism, and cylindrical wheel whereas puddler with pegs consisted of main body, *patta* or *patela* made of iron sheet with pegs, and driving mechanism. The effective depth of operation in both cases was 10cms, whereas the width covered was 1.14m in case of cage wheel puddler and 1.50m in case of peg puddler. The working capacities were 0.10 and 0.15 ha hr^{-1} for cage wheel and peg puddler, respectively. The estimated cost was Rs. 9000 approximately for both types.

9.4.1.7. Disc Harrow: Animal drawn disc harrow was usually single action type with single or double gang of disc. An off-set disc harrow permitted better coverage at the corners of a field or when working close to the trees in an orchard.

9.4.1.8. Spike Tooth Harrow: This was also known as Pegtooth harrow, drag harrow, section harrow or smoothening harrow. Its principal use was to smoothen and level the soil directly after ploughing. Normally, it penetrates about 5 cms. *Datari, Khutia* and a *Tikona* harrow were special cases of this type of harrow.

9.4.1.9. Spring Tooth Harrow: It was adapted for use in rough and strong ground. It was also used extensively to loosen soil in a previously ploughed field. The popularity of this type of harrow was limited.

9.4.1.10. Cultivator: Cultivator as a tillage tool tilled the field and stirred, aerated and mulched the soil. It also killed the weeds. Animal drawn cultivators are straddle type and inter-row type, with different kinds of cultivating shovels (reversible, single point and duck-foot shovels or sweeps-full and half, hillers, furrowers, etc.) e.g. Wah Wah 3- and 5-tynes cultivator, Damoh cultivator, Care Junior cultivator, etc.

9.4.1.11. Furrower: The furrower has been found useful for constructing open surface drainage channels required on deep black clay soils (Vertisols) under high rainfall conditions.

9.4.1.12. Rotavator: It had L-type blades mounted on the shaft. A good seed bed and pulverization of the soil was achieved in a single operation. It was also suitable for incorporation of straw and crop-residues into the soil.

9.4.2. Tractor drawn tillage implements

The bullock/animal power was being replaced by a tractor gradually. Tractor drawn tillage implements were being used not only by farmers with bigger holdings but even by farmers with medium and small size holdings, thanks to custom hire service now getting popular in villages, e.g. mouldboard plough, disc plough, cultivators and harrows were the tractor drawn tillage implements in common use.

9.4.2.1. Tractor driven Mouldboard Plough: it was used for deep ploughing to eradicate perennial obnoxious weeds such as *kans* (*Saccharum*), *jharberi* (*Zizyphus*) etc. It was one-way or two-way (reversible) plough. Also, depending upon the draw-bar horsepower of the tractor it could have more than one bottoms. A 35 HP tractor could pull 35cm three bottom plough. The plough should have accessories like rolling coulter jointer and gauge wheel (specially in case of trailing type units). Performance of mouldboard plough at high speeds was greatly aided by rolling coulter and jointer when considerable trash or surface weeds were present. If tall weeds were standing in the field, weed hooks or weed rods were helpful accessories of a mouldboard plough.

9.4.2.2. Tractor drawn Disc Plough: The disc plough was developed to reduce friction by making a rolling plough bottom instead of a sliding bottom as in mouldboard plough. Since disc plough does not create friction for penetration, it depends on its weight and disc angle for penetration. It was, therefore, heavily built. Its popularity was primarily due to its ability to work under adverse field conditions and low maintenance cost. It worked well in wet, sticky, hard, non-scouring soil. It rode over the stones and rocks rather than hooking on to them. It also rode over roots and stumps in the field by cutting these up. In very abrasive soils, the discs lasted much longer than the shares of a plough. It was well suited for deep ploughing. Further, mounted type disc plough was more popular than trailing type. Under normal field conditions, plough with 3 and 5 discs could be pulled easily by 35 HP and 50 HP tractors, respectively.

9.4.2.3 Chisel Plough (Sub-soiler): As the name indicates, it ripped its way through the soil with narrow double ended shovels (or chisel point) that are mounted on long shanks. Normally it penetrated in to the soil up to 50 cms depth. The soil was broken by tearing action of the plough. It was not inverted and pulverized. It was often used to loosen hard dry soil before using regular plough. It was also useful in breaking hard pan or plough sole. A 35 HP tractor could pull a single point chisel plough.

9.4.2.4. Mole Plough: It was a type of sub-soiler like the chisel plough. It was used for providing subsurface drainage to the waterlogged deep black soil. A 75 HP tractor could drive it about 60 cms below soil surface to make a pipeless circular channel to serve as a drain. The durability of this drain depended on physical condition of the soil and the climate. It can be used up to one month after termination of monsoon. The soil should be in plastic condition/state to enable construction of drainage channel at a desired depth. It costs Rs. 18000.

***9.4.2.5. Sub-soiler Plough*:** A sub-soiler plough or sub-soiler was built heavier than a chisel plough, since it was meant for ploughing deeper i.e. 50-90cms deep. A tractor of 60 to 85HP was required to pull a single ripper penetrating into hard soil at a depth of 90cms. The ripper of sub-soiler was usually long and thin with a heavy wedge like point.

9.4.2.6. Rotary Plough: It was also known as rotary tiller or rotavator. The rotor of the plough was mounted with knives, which pulverise the soil. The degree of pulverization obtained in a single run was equivalent to a tilth achieved by one ploughing, two disc harrowings and one run of a spike tooth harrow. Seed bed preparation by a rotary tiller saved time but the energy spent was more than a plough/harrow. Its main parts were: L-shaped blades mounted on a shaft, gear box, spurgear assembly, trailing board, etc. If deep ploughing was not to be done in a field, this machine could be used for puddling the soil to plant paddy seedlings. Its use saves 30-35 percent time, and performs 20-25 percent better quality work. Its effective width of operation was 150cms and depth 11 cms. Its estimated cost was Rs.1,25,000.

9.4.2.7. Disc Harrow: A harrow was used to crush clods, stir the soil and destroy the weeds. A disc harrow was used before ploughing to cut off vegetative material which could interfere with ploughing operation. A heavy disc harrow was used for summer cultivation. Tractor drawn harrow was double action type. Double action offset disc harrow was popular for straight field discing operation since it left the field smoother than a double action tandem disc harrow.

9.4.2.8. Spike-tooth Harrow: A spike-tooth harrow was also known as peg-tooth harrow. Its principal use was to smoothen and level the soil after ploughing. Normally it penetrated about 5cms.

9.4.2.9. Spring-tooth Harrow: Spring-tooth harrow was useful for stony ground, and was extensively used to loosen previously ploughed soil prior to sowing. Also, it penetrates deeper than the spike-tooth harrow, but its popularity was limited.

9.4.2.10. Cultivator: Cultivator as a tillage tool was used for tilling the field, stirring and pulverizing the soil. Tractor drawn cultivators were available with different types of cultivating shovels (reversible, single point, duck foot shovels or sweeps-full and half, etc.). A 9-tyne cultivator (9-tyne tiller) with rigid or spring loaded tynes was popular for hooking with 35 HP tractor. Duck-foot cultivator was found very suitable at Indore and could be adopted over large areas on black soils for shallow cultivation.

9.5. SEEDING AND PLANTING IMPLEMENTS

9.5.1. Hand Dibbler

Hand dibbler was used to sow seeds of big or medium size on well-prepared seed-bed. It was also used to fill in gaps in plant rows due to defective germination which was a big and usual problem in rainfed farming. There were two types of hand dibblers, viz. (i) Navin Dibbler and (ii) Rotary Dibbler. The Navin Dibbler was for sowing seeds deeper (35mm) as compared to rotary Dibbler (20mm). The man-hours ha^{-1} required were also 36 and 22 for Navin and Rotary dibbler, respectively. The Navin dibbler had jaw-type seed-dropping mechanism, wooden measuring device, lever type transmission system for jaw, seed box, handle and seed outlet. Its cost was Rs. 700. The Rotary Dibbler consisted of seed hopper, seed outlet jaw, driving lever jaw, wooden roller, handle, transmission and soil compacting wheel, etc. Its cost was about Rs. 2300 (Dube and Singh, 2014).

9.5.2. Man driven Paddy seeder

This planter was used to sow germinated paddy seeds on puddled field. The implement was available in 4 rows to 8 rows planter size, with inter-row spacing of 20cms. The seeds were put in a rotating drum with an adjustable hole, so that seeds fell in rows. The cost of the implement was Rs. 3600 and Rs. 5400 for 4-row and 8-row planters, respectively.

9.5.3. Hand operated Paddy transplanter

Paddy transplanter was used to transplant paddy seedlings into puddled field. The man driven machine was a six-row planter consisting of a handle, seedling holder, seedling tray, transplanting system, main body and wooden platform. The row to row spacing was 20 cms, operational width 120cms and depth of planting 2 to 3 cms. Its cost was Rs. 12,000. The self propelled 8 row machine had an engine of 55HP and cracking type transplanting system to adjust plant to plant distance. The row to row spacing was 24cms and the working width 192cms. The machine costed Rs. 4,50,000.

9.6. BULLOCK DRAWN SEEDING EQUIPMENT

Timely and precise placement of seed and fertilizer was necessary to obtain an adequate plant-stand which was the greatest problem of dry farming. Placement of fertilizer (particularly phosphatic and potassic fertilizers) below seed was found to be most effective method of fertilizer application under dryfarming: Accordingly, various seed-cum-fertilizer drills have been tried, while Deshi plough with seeding attachment was still in use (Fig. 23).

Fig. 23. Deshi Plough with Seed and Fertilizer placement Attachments

9.6.1. Seed Drills

Of various seed-drills tried earlier, *Kenmore* Tool Bar with *Siser* seeder, *Agricart* tool bar with *Ebra* seeder, and M.P. seed drill were found promising while *Mahakal Duphan* and *Mahakal Tiphan* became very popular amongst the farmers particularly in Malwa Region during 1980s. Later on, a number of different types of seed drills became available for sowing soybean, sorghum, wheat, chickpea, and safflower. In a comparative six year (1990-95) study under All India Coordinated Research Project on Dryland Agriculture Centre Indore, the performance of seed and fertilizer placement, draft, etc. of five seed-cum-fertilizer drills viz. *Malviya* seed drill, CIAE seed drill, (designed by Central Institute of Agriculture Engineering, Bhopal), *Mahakal Tifan* (3 row planter), *Mahakal Duphan* (2 row planter) and *Shivaji* seed drill were tried. CIAE seed drill was found to be the best with Operational Performance Index (O P I) of 0.85, followed by *Malviya* Seed drill with OPI of 0.83. (Jain et al 1998).

9.6.2. NRC Seed Drill

Recently the National Research Centre for Soybean (NRCS) I.C.A.R., Indore, had designed one improved seed-drill for soybean based cropping system. The seed-drill was used successfully by a large number of farmers on custom-hire basis.

9.6.3. Tilplant machine

To ensure double cropping under rainfed conditions when there was adequate moisture content in soil at the time of harvest of *kharif* crops, it was necessary to sow the *rabi* crop as soon as possible without even the preparatory tillage. It would enable not only early sowing but would also assure the availability of maximum amount of soil moisture for germination and growth of the crop plants. For this, a 'Tilplant' machine was needed. Accordingly, an effort to develop such a machine was made by the Indo-UK Dry Farming Operational Research Project, JNKVV (now RVSKVV) Campus Indore, during 1977-78. A prototype with duck foot type tiller and the seeding machine–two in one type gadget, was fabricated and tested successfully on farmers' fields. Further research work on it was conducted at College of Agricultural Engineering, JNKVV, Jabalpur, and bullock drawn Jabalpur Till Planter was designed (Alam 1994). But unfortunately the programme of its manufacture was not undertaken. However, a tractor mounted Till–plant machine was now available.

9.6.4. Bullock drawn Planter

The planter is 3-row inclined plate planter for sowing both bold and small seeds, and for sowing intercrops. It consists of a frame with tool bar, modular seed boxes, furrow openers and ground wheel. Seed plates for sowing different seeds can be selected and easily changed. Shoe type furrow openers ensure deeper seed placement in moist zone for sowing under dryland conditions. Furrow opener units can be adjusted for different row to row spacings.

9.6.5. Bullock drawn Semi-automatic Ridge Planter-cum-Fertilizer Drill

The machine was developed at College of Agricultural Engineering, JNKVV, Jabalpur, by Dr. Alam and Ray in 1978. It formed ridges in pulverized seedbed and drilled seed and fertilizer in one pass. Adopting bund-former it has a hopper with gravity feed metering with agitator and spoked ground drive wheel mounted directly on metering shaft, a shut-off lever for fertilizer, and a seeding funnel and tube. It was tested and found satisfactory for ridge planting/drilling of pigeonpea, blackgram, and greengram.

9.7. TRACTOR MOUNTED ZERO-TILL SEED-CUM-FERTILIZER DRILL

Zero tillage was being emphasized and popularized amongst the farmers in a big way by Rice-Wheat Consortium for the Indo-Gangetic Plains (RWC-CIMMYT). This gave a great fillip to development of zero-till seed-cum-fertilizer drill and the National Agricultural Research and Extension system developed several versions of the Zero-Till Seed- cum- Fertilizer Drill and Zero-Till Drill-cum-bed Planter proto types. These proto types are being improved continuously by manufacturers (Yadav et al 2002) with active involvements of the national scientists using the feedback from user farmers (Figure 24). It consisted of a frame, seed-box, fertilizer box, seed metering mechanism, fertilizer metering mechanism, furrow openers depth-control side wheels, hitch points, iron/wooden platform or stand and transport-cum-power transmitting wheel. The frame was made of mild steel and tynes were mounted with clamps to obtain variable row spacing. Furrow openers were inverted T-type with narrow shovels to penetrate the soil easily. The main advantage of narrow shovels was reduced penetration force and lower draft requirement.

Fig. 24. Zero-till seed cum fertilizer drill

9.7.1. Tractor mounted ridger seeder

The implement was used for making ridge and furrows and to carry out sowing of seeds either on the ridge or in the furrow. The seeder consists of seed hopper, fertilizer hopper, seed metering mechanism, fertilizer metering mechanism, drive wheels, hitch system, furrow openers and sturdy frame made of mild steel sections. The ridger body consists of two mouldboards and share mounted in the front. The wingspan of the ridger can be adjusted to vary width of furrow/ridge. Seed and fertilizer metering units received power from ground wheel through chain and sprocket system.

9.8. INTER-CULTURE TOOLS AND EQUIPMENT

Inter-culture operations in standing crops specially *kharif* ones were necessary to eliminate weeds and increase soil aeration particularly on black soils. Both hand operated tools and bullock drawn equipment had been in use for inter culture operations. A preliminary study on the field performance evaluation of locally manufactured inter culture tools namely hand hoe, bullock drawn *dora* (mini blade harrow), rotary hoe, *Jalo* cultivator, wheel hoe and Indore ridger conducted under Indo-UK Dry Farming Project, Indore, to compare their power to destroy weeds, area coverage, and ease of handling revealed that bullock drawn *dora* and 'Indore ridger' were superior and already popular among Malwa and Nimar farmers. These were very promising for adoption over large areas under black soils.

9.8.1. Bullock drawn Ridger Cultivator

It was developed at College of Agril Engg, JNKVV, Jabalpur, M.P. by Dr. A. Alam and Utpal Ray in 1977, for interculture and intra-row weeding by smothering with soft scraped soil thrown towards the rows. On a double handled frame it had a sweep in the middle and two gangs of scraping blades on a special purpose standard on either side of the sweep. Front scraping blades scraped top rough soil pushing it to the middle of the row and rear blades scraped soft soil and threw it towards the plants covering young emerging weeds. It was suitable for inter-culture in wide row sown crops in friable soil condition.

9.8.2. Earthing Attachment to '*Dora*' (Blade harrow for inter-culture)

It was developed at College of Agril Engg, JNKVV, Jabalpur, M.P. by A. Alam and Utpal Ray in 1977. It converted '*Dora*'-a miniature bullock drawn blade-harrow, into an inter-culture-cum-earthing device. It was an improvised furrower snap-on mounted on a conventional '*Dora*', suitable for wide row sown crops.

9.8.3. M.P. Double *Dora*

Recently a bullock drawn M.P. Double *Dora* with two mini-blades attached to a single beam was getting popular. The height of the beam could be adjusted to suit the height of the bullocks. The shape of the blade was slightly concave. These blades could be replaced by V-shaped blades in case of narrow spaced rows of crops like soybean. V-shaped blades did a little earthing besides scraping the soil.

9.8.4. Wheel Hoe

Wheel hoe was a manually operated, widely accepted tool for weeding and inter-culture operation in row-crops. It had a long handle and was operated by pull and push action. It consisted of a wheel assembly, miniature tool frame and replaceable blades such as straight blade, reversible blade, sweeps, V-blade, tine cultivator, miniature furrower and spike harrow. These are made of hardened carbon steel. The working depth of the tool and the handle height can be adjusted so that it could be operated by a single person.

9.8.5. Double Wheel Hoe

This was a two-wheeled manually operated weeder. It consisted of a shaft with two wheels, V-blade, U-clamps and of long handle. It was also operated by push-and pull action to cut and pull away weeds.

9.8.6. Power Tiller Operated Inter-culture Machine

The machine consists of main body, movement wheel, handle, driving mechanism and sweep type blades. The machine was propelled by power Tiller of 8-10HP. The machine could be operated by a single person in widely spaced rows of a crop. Its effective width of operation was 135 cms and depth 5cms. It covered 0.18 to 0.25 ha hr^{-1} and costed about Rs. 8000.

9.8.7. Rice cono-weeder

The weeder consisted of two rotors, float and a handle joined to the frame. The rotors are made of smooth, serrated strips welded to the surfaces along their lengths. The float controlled the working depth and did not allow rotor assembly to sink in the puddle. The weeder was operated by pushing action and stirred top 3 cm layer of soil.

9.9. HARVESTING TOOLS

9.9.1. Sickle

Sickle was an indigenous tool used for manual harvesting of various crops. It was modified to increase its efficiency as serrated blade sickle.

9.9.2. Serrated blade sickle

The cutting edge of this sickle was serrated. It was slightly heavier and thicker than ordinary sickle. Made of mild steel, it consisted of blade with tangle and a wooden handle. It was used for harvesting paddy and other field crops and for intercultural operation also.

9.9.3. Naveen sickle

This is also a serrated blade sickle, made of medium carbon or alloy steel, tempered to suitable hardness. The wooden handle had a bend at the rear for better grip and to avoid injury to hand during operation. It was more efficient than local sickles.

9.10. ANIMAL AND TRACTOR DRAWN DIGGER

To dig out groundnut, potato, sweet potato, etc. from soil was a laborious job. Animal drawn digger consists of body, handle, drive wheel, depth adjustment system and V-blades. The V-blades are 30 to 45cms apart depending upon crops. The tractor drawn digger was bigger and heavier covering 135cms width of crop-rows, and operated by 25-30 HP tractor. Its working capacity was 0.4 ha hr^{-1}.

9.11. THRESHERS

After harvest of a crop, threshing was done to separate the grain from straw/ stover. Traditionally, the crops were harvested by men using sickles and threshed by man or bullock-power. Now, a power driven multi-crop thresher of different capacities and harvester-cum-threshers i.e. combines are available.

9.11.1. Axle Flow Multi-crop Thresher

In this machine, the harvested crop fed into it, moves over a cylinder towards axle and thus greater traction energy was available to separate grain from earhead. It was suitable for threshing of wheat, soybean, sorghum, maize, paddy, sunflower and safflower. This was powered by a 7.5 H.P. three-phase electric motor. It consisted of raspbar, palted cylinder, bottom, semicircular plate, blower, screen and rubber transmission wheels. A higher capacity machine could thresh wheat, maize, sorghum, pigeonpea, soybean, chickpea, sunflower, etc. and was operated by a 20 H.P. three-phase motor or by 35 H.P. tractor P.T.O. It consisted of a feeding spout spike tooth cylinder, three aspirator blower, hitch system and rubber transmission wheel. This machine was available on hire and very useful to farmers, particularly those growing pigeonpea with other crops. Its estimated cost was Rs. 100,000 (Dubey and Singh, 2014).

9.11.2. Minor-millets thresher

This machine was used to thresh minor millets such as kodo and kutki. It was powered by a 20H.P. motor. The main parts of the machine were: threshing unit, blower, screen (concave) and main structure. Its working capacity was 20 to 50 kg hr^{-1}, and 2 to 3 percent grains were broken. Its estimated cost was Rs. 80,000.

9.11.3. Combine Harvester

The machine was operated by 35 H.P. tractor P.T.O. It has a standard unit for collecting straw, standard cutter bar for cutting the plants (to be harvested), augar type feeding unit, threshing drum, blower, tray for collecting grain and so on. After threshing, the straw got stored in a box situated on the frame of straw reaper. When the straw-store was full, it was emptied at a suitable place with the help of hydraulic system. The machine was very convenient to operate and the whole process of harvesting, threshing and separation of grain and straw was completed easily and automatically. Also, less effort and space was required for the machine to turn in the field. The trolley was provided with a hood so that straw was dropped into it safely and was not blown away with the wind. The effective width of the machine was 240 cms and covered 0.37 ha of cropped area per hour. The estimated cost of the machine was Rs. 3,25,000.

9.12. OTHER EQUIPMENT

9.12.1. Groundnut Decorticator

A hand operated Groundnut Decorticator has been found ideal for small farmers. The machine is operated by two labourers and has the capacity of 100 kg hr^{-1} with breakage of kernals not exceeding 5 percent (Sharma 1980).

9.12.2. Low lift irrigation pump

For recycling the run-off collected water, a low lift, reciprocating and manually operated pump has been tested and found very satisfactory upto a suction of 2 metres. It has a capacity of 4000 to 5000 litres hr^{-1} and is operated by two labourers (Sharma 1980).

■■■■

X

INTEGRATED FARMING SYSTEMS FOR LIVELIHOOD SECURITY

Rainfed farming in India was still subsistence enterprise and suffered from twin problems of low and unstable productivity. Since crop-productivity under rainfed conditions suffered from vagaries of monsoon and other climatic calamities like frost, untimely hot and dry winds particularly at flowering stage of crops, it was necessary to have a second string to farmers' income. This could be easily achieved by adopting mixed farming instead of arable farming. Mixed farming was not only more remunerative but it also provided greater employment to family labour and extra yields of grain, fodder and cash crops (Kulkarni and Chavan 1980). Studies carried out in the dry farming areas all over the country indicated that the income through raising fodder crops for rearing animals and sustaining milch animals was higher compared to the returns obtained from raising dryland cereal crops. The net return per hectare was 26 per cent higher in case of fodder crops as against the dry land crops (Reddy and Sampath 1978). Mixed farming was economically and socially viable under rainfed conditions particularly in drought prone areas. There was lot of unproductive land (shallow, coarse textured soil with almost nil water holding capacity) in villages which was not suitable for growing crops but could sustain remunerative livestock farming. Also, there may not be any production from crops grown on such land in a drought year, but income from livestock farming may not be affected and this might serve as an insurance against the drought. While there was lot of scope to improve this traditional mixed farming-keeping milch animals along with growing of crops, there were other integrated farming systems for different agroclimatic conditions to be adopted by small and marginal farmers and landless labourers for their livelihood security. A number of these integrated farming systems have been tested on farmer/landless persons' households under National Agricultural Innovation Project, (NAIP during 2006-2013) and found to be much more remunerative/generating high pay-offs, for their livelihood security, e.g. Rice-fish-poultry farming system, Integrated rice-fish-vegetable system, Integrated fish-vegetable-poultry/pig system, Fish-livestock-horti systems, Fish-duck-pig based farming system, Fish-*makhana*/water chestnut integrated farming system, Silvi-pastoral system, Agri-horti system, Horti-pastoral system, Mushroom production, Poultry (Kadaknath) farming, Goat farming, Tsar sericulture, Lac cultivation, and so on. Thus, besides improving traditional systems, need for adoption of new integrated farming systems appropriate for local conditions, was apparent.

10.1. TRADITIONAL MIXED FARMING

As already stated, raising of milch animals (cows and buffaloes) with crop-production was commonly followed by farmers. But the productivity of cows/buffaloes with the farmers was very low. The average milk production (litres per day) per non-descript cow, crossed cow and buffalo, for the whole Nation was 1.970, 6.437 and 4.300, respectively, while the average figures for Central India were 1.759, 5.992, 3.334, respectively. Thus, the productivity of the cow and buffalo with the farmers in general, was extremely low. However, the productivity of cross-bred cow was more than three times as compared to that of scrub cow. Therefore, the mixed farming could be much more profitable, if the genetic quality of milch animal was upgraded. Main reasons for the low milk yield per animal were: (1) scrub cattles and buffaloes, (2) long dry periods, (3) high death rate of calves, (4) improper and inadequate feeding and management, (5) low grade pastures, and (6) inadequate veterinary services. Therefore, the important practices for improving the productivity of livestock are: (a) improvement of feeding and management, (b) immunization against important diseases, and (c) upgrading of scrub cattle.

10.1.1. Improvement of feeding and management

The present level of feeding and management of live-stock was very low and the same had to be improved considerably. For this, production of fodder and forage crops in rotation with grain crops was necessary. Also, improvement of pastures had to be undertaken.

10.1.1.1. Production of fodder and forage crops: The livestock population of India increased from 291.9 million in 1951 to 485 million in 2003 (Table 58). Of this total livestock population (485 million), the largest population was of cattle (185 million) followed by that of goat (124 million). While the number of goats increased at the fastest rate (3.3 percent per year), surprisingly the rate of increase of buffaloes was the next and not that of cattle. The population of buffaloes increased from 43.3 million in 1951 to 97.9 million in 2003 (2.5 percent per year), while that of cattle from 155.3 million to only 185.2 million (0.037 percent per year) during the same period. Due to large increase in livestock population, there was an unusually high pressure on pastures and grazing land and consequently their productivity was reduced to minimum. As a matter of fact, grazing land could provide some support to feeding of cattle and buffaloes during monsoon months only but for the remaining period of the year, the animals had to be maintained on crop-residues, straw or grown fodder crops. Apparently, there was an urgent need to increase the production of fodder and forage crops. Fortunately, there was a big scope to increase the productivity of these, both horizontally and vertically.

Table 58. Livestock population (1951-2003) of India (millions) (IASRI and ICAR 2005)

Livestock	1951	1961	1972	1982	1992	2003
Cattle	155.3	175.6	178.3	190.8	204.8	185.2
Buffaloes	43.3	51.2	57.5	69.0	84.2	97.9
Sheep	38.4	40.0	42.0	48.1	50.8	61.5
Goat	47.1	60.9	67.5	94.7	115.3	124.4
Others*	7.8	8.5	10.0	13.3	15.8	16.0
Total	291.9	336.2	355.3	415.9	470.9	485.0

*Others include horses, ponies, pigs, camels, donkeys, mules, yaks and mithun.

10.1.1.2. Horizontal increase in productivity: Cultivated fodder and forage crops were usually the cheapest source of feed nutrients required by farm animals. But farmers, in general did not set even a small part of their holding for growing fodder crops, except of course the dairy farmers. It was necessary to educate and convince the farmers about the urgency of growing fodder for remunerative livestock enterprise. Also, it was necessary to introduce fodder crops in prevailing cropping systems practiced by the farmers. Further, a fodder crop of sorghum or maize or teosinte (makchari) could be raised over large areas which were left fallow at present. Utilization of upland fallows for cultivation of maize to provide fodder and grain increased income, food and nutritional security of tribals in Bastar region of Chhattisgarh (NAIP 2012). Similarly cultivation of sorghum can be profitably practised over extensive areas under shallow and medium black soils in semi-arid region. Being of short duration a fodder crop would not disturb the farmers' normal *rabi* programme. The fodder crop could be harvested much earlier than normal *kharif* crop and consequently there should be enough moisture in soil as well as the time required for preparation of seed-bed for rainfed *rabi* crop, after the harvest of fodder crop. This would also avoid over tilling of soil during monsoon season, usually practised by farmers to get a very fine seed bed for *rabi* crops, particularly wheat and mustard. The fodder crop would also provide a good vegetative cover to the land which was otherwise exposed to enormous soil erosion during the monsoon, particularly under open field cultivation system prevalent in western M.P., Vindhyan region, and Narmada Valley. Further, in low rainfall (arid) and drought prone areas, it was safer to adopt livestock farming than arable farming by growing fodder and forage crops to feed the animals. Also, it was more remunerative to grow grasses and fodder shrubs on marginal and degraded slopy lands rather than put these to arable farming. Fodder and forage crops played very important role in appropriate management of such lands. Forage crops could also be raised as intercrop strips across the slope between arable crops, as a soil conservation measure. Thus, there was great scope to increase production of fodder and forage crops in vast areas not appropriately used/managed at present.

10.1.1.3. Vertical increase in productivity: There was tremendous scope of increasing the production of fodder and forage crops by increasing their yields per unit area. The yield of fodder crops could be increased considerably through good quality of seed, selection of varieties and adoption of improved package of practices.

10.1.1.3.1. Seed: The biggest constraint in popularizing the fodder and forage crops and in getting their high yields per unit area was the lack of availability of adequate quantities of quality seeds. As both fodder and forage crops were shy seeders, the availability of their seeds posed a serious problem. Even at present there is no organized agency for the production, certification and distribution of fodder seeds in the country and in the states, although there are several multinational companies and semi-government, autonomous seed corporations (both National Seed Corporation and State Seed Corporations) dealing with vegetable seeds and field crop seeds. Thus, the need for an organized agency for the production and certification as well as for distribution of fodder seeds could not be over-emphasized. Also, special fodder seed production programme should be taken up at Government farms, Military Dairy Farms and private dairy farms under the guidance of experts. Indian Grassland and Fodder Research Institute (IGFRI), Jhansi has conducted commendable work on various cropping systems (mixed cropping, intercropping, alley cropping etc.) on varieties of different fodder crops, compatibility of different fodder and forage crops with different fodder shrubs/trees and other aspects to augment the vertical productivity of these crops. However, the technology was not reaching the farmers and would not reach them unless strong seed production programme for these crops was undertaken as suggested earlier.

10.1.1.3.2. Fodder Crops and Varieties: There were several varieties of fodder sorghum and maize suitable for growing alone or in combination with cowpea. Teosinte (makchari) which was a cross between pearlmillet and maize had also been found to be a very good fodder crop. It was much sweeter as compared to maize and was, therefore, relished and consumed very greedily by animals. Besides sorghum and maize, other fodder crops are pearlmillet, cluster bean, cowpea, greengram, and velvet bean suitable for arid and semi-arid regions. In *rabi*, lucerne and berseem, with or without oat, could be planted for fodder where irrigation by well or run-off collected water was available. Suitable varieties of these crops recommended for different regions were given in table 59 (Hazra 1995).

10.1.1.3.3. Forage crops and Varieties: As already pointed out, forage crops could be grown on marginal and sub marginal lands in arid and semi-arid regions of the country. Various grasses not only yielded nutritious fodder for livestock but also helped soil-stabilization against erosion. Species of grasses e.g. *Pennisetum pedicellatum, Cenchrus setigerus, Cenchrus ciliaris, Sehima sulcatum, Chrysopogon fulvus, Heterapogon contortus, Dicanthium annulatum, Bothriochloa pertusa* and *Iseilema laxum* with legumes such as *Macroptelium atropurpureous, Alysicarpus monilifer, Indigofera linifolia, Rhynchosia minima* and *Stilozobium dearingianum* were very suitable for marginal and ravineous (shallow ravines) lands. Grass species and their varieties recommended for different regions of the country were presented in Table 60 (Hazra 1995).

Table 59. Recommended varieties of fodder crops for different regions

S.No.	Crop	Varieties	Region for which recommended
1.	Sorghum	SL-44, MP chari	North India
	a. Single cut	SL-136, Raj chari-1	Whole Country
	b. Multi cut	AS-16, Meethi sudan (SSG-59-3), Hara sona (855 F)	Whole Country
2.	Pearlmillet		
	a. Single cut	UUJ-IV-M, TNSC-1	Entire growing tract
	b. Multi cut	Giant bajra	Central India
3.	Maize	Vijay composit, J-1006	Whole Country
4.	Teosinte	TL-1	Whole Country
5.	Cowpea	EC-4216, IFC-8503	North, West and Central India
		UPC-287, UPC-5286 IFC-8401	Whole country
6.	Clusterbean	FS-227, HFG-119, IGFRI-212, IGFRI-2395-2	Entire growing tract
7.	Berseem	JB-1, JB-2, JB-3	Central India
8.	Lucerne		
	a. Perennial	T-9	Entire growing area
	b. Annual	LIC-3	Entire growing area

Table 60. Grass species/varieties for different regions

S. No.	Crop	Varieties	Region for which recommended
1.	Napier-bajra hybrid	Pusa giant, NB-21	Whole India
		IGFRI-3, IGFRI-7	Central and North-East India
		IGFRI-10	Whole Country
2.	Guinea grass	Hamil	North, West, Central and South India
		PGG-1 and PGG-9	North, North West and Central India
		PGG-14	Central India and Hills
3.	Dinanath grass	PS-3, IGFRI-4-2-1 and IGFRI-4-3-1	Whole Country
4.	Setaria	Nandi, Narok	Whole Country
5.	Anjan grass	Bundel Anjan (IGFRI-3108) Marwar Anjan (CAZRI-75)	Arid and Semi-arid Region
6.	Blue Anjan	CO-1 (FS-391)	Semi-arid area
7.	Yellow Anjan	Marwar Dhaman (CAZRI-76)	Arid and Semi-arid Regions
8.	Marvel grass	GMG-1	Central and Western India

10.1.1.3.4. Package of practices: For full exploitation of yield-potential of recommended varieties, it was necessary to follow improved package of practices. Rain-water management, adoption of integrated plant nutrient management based on soil-test (application of organic manures such as farm yard manure (FYM), compost, vermi-compost, non-edible oil cakes, waste from sugar mills (pressmud), urinated litter from animal sheds, bio-fertilizers particularly rhizobium cultures for legumes, phosphate solubilising bacteria, etc. supplemented by fertilizers, integrated pest management (use of chemicals only if necessary) and harvesting at appropriate time, were all necessary practices to get good quantity and quality fodder. Inter culture operations were not required in fodder crops except up-rooting of obnoxious plants and detrimental weeds. Tripathi and Agrawal (2003) emphasized the role of bio-fertilizers in intensive fodder production and brought out that application of Azotobacter/Azospirillum to different cereal crops and grasses resulted in saving of 20 to 25 kg of nitrogen ha^{-1}. Also, Rhizobium inoculation of legume forage increased their yield from 14 to 46 percent. Application of VAM-fungi to guinea grass resulted in saving of 25 kg nitrogen ha^{-1}. Inoculation of berseem seed with phosphate solubilising bacteria increased the green fodder yield (103.6 tonnes ha^{-1}), dry matter yield (16.19 tonnes ha^{-1}) and crude protein yield (3.30 tonnes ha^{-1}). Efficient use of farm yard manure and vermicompost (each 5 tonnes ha^{-1}) yielded 40 to 42 tonnes ha^{-1} of green oats fodder.

10.1.1.4. Fodder shrubs and trees: Many shrubs and trees provide top feed to animals in arid and semi-arid zones and in drought prone areas, e.g. species of Acacia (*Acacia nilotica, A. catechu, A. tortilis*) and Prosopis (*P. cineraria, P. juliflora*), trees like *Azadirachta indica, Albizia lebbek, Dalbergia sissoo, Hardwickia binata, Ficus bengalensis, Ficus religiosa, Bauhimia variegate* and *Butea monosperma*, and shrubs such as *Leucaena leucocephala, Zizyphus nummularia, Z. rotundus* and *Z. maurisiana.* Plantation of these trees and shrubs on farm boundaries, field bunds, road sides, village surroundings and community grazing land could go a long way to overcome fodder shortages. Trees recommended for planting under water-logged conditions are safeda, poplar (*Populus alba*), arjun (*Terminalia arjuna*), babool (*Acacia nilotica*), jamun (*Eugenia jambolana*), safed siras (*Albizia procera*), shisham (*Dalbergia sissoo*), khejri (*Prosopis cinereria*), karanj (*Pongamia pinnata*), etc. (Koshta 1997).

10.1.1.5. Alley cropping: Alley cropping is a kind of intercropping to increase the production per unit area by judicious combination of crops and trees/bushes or grasses and trees/bushes. Alley cropping was very beneficial because it could provide fodder and fuel besides regular production of grain and straw from crops, and it yielded useful products even in drought years. The choice of food crops, grasses, bushes and trees for alley cropping system would depend upon: (i) need of the farmer, (ii) type/quality of his land and soil, and (iii) climatic conditions of the area. These factors would determine whether an agro-forestry or a silvipastoral or a hortipastoral system was needed. For forage and fodder production, following combinations are suggested:

i. Dinanath grass (*Pennisetum pedicellatum*) with Subabool (*Leucaena leucocephala*)
ii. Cowpea/Cluster bean with Castor/Eucalyptus/subabool
iii. Sorghum/Maize/Pearl millet with Stilo/Siratro/Anjan grass
iv. Albizia species/Acacia species of trees with Cenchrus species/Sehima species of grasses.

To get maximum return from alley cropping, the hedges of subabool, eucalyptus, etc. should be logged/pruned to prevent shading effect on alley crops.

10.1.1.6. Other sources of fodder (Non conventional): According to Mishra et al (2003), about 35 million tonnes of non-conventional straws from crops such as soybean, safflower, rapeseed/mustard, taramira (*Eruca sativa*), cluster bean, green gram, black gram etc. was not utilized every year. These are fibrous and hard materials and have low digestibility and low palatability due to their high content of ligno-cellulosic and silica content. These can be chaffed and mixed with some normal fodder/forages to feed cattle not in lactation.

10.1.1.7. Mixed cropping: Mixed cropping of fodder and forage crops with legumes are more advantageous than growing single crops. Various advantages were:

i. Higher yield from mixed cropping due to synergistic effect of legumes
ii. More nutritive, succulent and palatable fodder
iii. Much greater quantities of digestible dry matter and protein, and
iv. Improvement of the physical condition and fertility status of soil by providing better vegetative cover to soil during monsoon season and leaving of nitrogen-rich residues by leguminous crops.

Some common crop-mixtures are

1. Sorghum (*Sorghum vulgare*) + cow pea (*Vigna unguiculata*)
2. Sorghum (*Sorghum vulgare*) + clusterbean (*Cyamopsis tetragonoloba*)
3. Maize (*Zea mays*) + cow pea (*Vigna unguiculata*)
4. Pearl millet (*Pennisetum typhoides*) + cluster bean (*Cyamopsis tetra gonoloba*)
5. Dinanath grass (*Pennisetum pedicellatum*) + Stilo (*Stylosanthes hamata*)
6. Napier grass (*Pennisetum purpureum*) + Siratro (*Macroptilium atropurpureum*)
7. Anjan grass (*Cenchrus ciliaris*) + Subabool (*Leucaena leucocephala*)

10.1.1.8. Supplementation of diet with mineral mixture: Based on research conducted in Jawaharlal Nehru Krishi Vishwa Vidyalaya, Jabalpur, (Raghu et al 2008) supplementation of diet with mineral mixture @ 30 g per day per animal from calving to fourth month of lactation for cattle, and @ 40 g per day per animal from calving to fourth month of lactation for buffalo was recommended.

10.1.2. Improvement of pastures and grazing lands

Most of the pastures and grazing lands whether individually or community owned, were over grazed, i.e. grazing intensity was very high as compared to optimum carrying capacity. Also, these were misutilised and were therefore in a pitiful condition. Most of these served as exercise grounds for the animals rather than grazing grounds. These needed rejuvenation and scientific management by planting better species of grasses and legumes. In arid and semi-arid regions, various grasses like guinea grass (*Panicum maximum*), blue panic (*Panicum antidotale*), napier grass (*Pennisetum purpureum*), anjan grass (*Cenchrus ciliaris*), pila anjan (*Cenchrus setigerus*), marvel (*Dicanthium annulatum*), ponia (*Sehima nervosum*), lemon grass (*Cymbopogon flexuosus*) and broom grass (*Thysanolaena maxima*) and legume species such as stilo (*Stylosanthes hamata, S. scabra*), siratro (*Phaseolus atropurpureum*), pilli pasara (*P. trilobus*), chipkani (*Desmodium contortum*), sheora (*Alysicarpus rugosus*), and subabool (*Leucaena leucocephala*), have proved valuable. At the time of planting grass, it was necessary to apply organic manures and fertilizer based on soil test. Also, it was necessary to top dress urea/ammonium sulphate after grazing/cutting of the fodder initially. Later on, the symbiotic effect of legumes associated with grasses would take care of the nitrogen needs of grasses.

10.1.2.1. Problems of pasture management: The biggest problem in managing the pastures after introducing high yielding grasses and legumes was overgrazing and the damage done to pastures by trampling of the plants under the feet of grazing animals. That was why the forage biomass increased many fold when the pasture/grazing land was protected against free grazing. Not only this, the ratio of palatable species also remained intact under protection against grazing. Therefore, to maintain the productivity of grazing land, following steps were necessary:

i. Controlled grazing through rotational grazing, regulating the number of animals according to carrying capacity of the grazing land on part of it at a time,
ii. Introduction of cut and carry system for stall feeding of animals instead of free grazing by animals, particularly when the soil under the pasture was wet.
iii. Application of organic manures and fertilizers: The soil under common grazing land/pastures was impoverished by continuous over grazing pressure and no one's care to maintain soil fertility. Consequently, even a small dose of fertilizers increased the productivity of the land many times. Based on soil testing, appropriate quantities of FYM, compost, vermi-compost, bio-fertilizers, non-edible oil cakes supplemented by fertilizers had to be applied.

10.1.2.2. Transfer of technology: While technology for improvement and appropriate management of pastures and grazing lands was well established, its transfer to beneficiaries was a big problem. First of all, farmers need to be convinced that for degraded and problematic land (slopy eroded, ravineous, saline-alkali land) silvipastoral system was the most appropriate for sustainable livelihood, keeping livestock as the basic component. For this system of living on dairying, farmers have to be organized as Cooperative Society/ Self Help Groups to improve and manage community grazing lands/pastures so as to avoid over grazing and spoiling the grazing lands by free grazing particularly during rainy season. Only through participation of such groups, practices like rotational grazing and utilising forage by cut-and-carry method can be got adopted by the beneficiaries. Not only this, in such organized forms they can demand better price for their products-milk and milk products, and take advantage of government subsidy and other welfare schemes much better than as individuals, who regularly migrate to towns in search of work for cash. This group of small farmers, with marginal land and landless labourers could also undertake cooperative mixed farming and embark upon intensive and improved livestock farming by pooling their resources.

10.1.3. Disease and Parasites control

10.1.3.1. Immunization against diseases: Besides improvement in feeding of the livestock, it was necessary to protect them from diseases. Some infectious diseases take a heavy toll of good animals every year. These can be easily prevented by immunizing the animals against these diseases e.g. Haemorrhagic septicemia, black quarter and rinderpest.

10.1.3.2. Parasitic infections: Helminthus nematode was the major parasite infecting animals. The time tested treatment against the nematode infection is use of anti-parasitic drugs called anthelmintics which destroy worms. In the past few decades, the extensive and indiscriminate use of anthelmintics had resulted in wide–spread parasitic resistance. Accordingly plant-based anthelmintics have been discovered and recommended e.g. powder of Palas (*Butea frondosa*) seeds, dry leaves of Neem (*Azadirachta indica*), and hot-cold extract of Papaya (*Carica papaya*). The infected animals should be isolated and their excreta collected and disposed off properly as it could be a source of infection (Raghu et al 2008).

10.1.3.3. Wounds and inflammatory disorders: These are very common in animals and antibiotics are usually prescribed for their treatment. Following medicinal plant preparations were recently recommended for wound healing:

i. Leaves of adusa (*Adhatoda vasica*) or nigundi (*Vitex negundo*) dried and powdered and mixed with sufficient vaseline to prepare ointment for application.
ii. Roots of ashwagandha (*Withania somnifera*) dried and powdered and mixed with 5 parts of vaseline to make an ointment to apply to the wounds.

10.1.3.4. Deworming: First deworming of buffalo calves with piperazine @ 30 ml per calf was recommended at the age of one week, to be repeated after three weeks.

10.1.4. Upgradation of scrub cattle and buffaloes

After the feeding and management has been improved, the scrub cattle and buffaloes should be upgraded using artificial insemination or direct service methods whatsoever at hand. It was controversial whether upgradation by crossing with pure Indian breeds or by exotic ones was more profitable. As improvement of local cattle through selective breeding was a very slow process, cross-breeding local cattle with exotic breed should be adopted as the national policy for the improvement of cattle (Parikh 1998). The results of cross-breeding at the organized farms had been very encouraging. First generation cross-breds had given a lactation yield of more than 3000 litres. However, the performance levels of cross-breds in rural areas did not appear to be very encouraging, and the same was almost 50 percent of the performance of the cross-bred at the organized farms. However, the performance of different cross-breds with exotic males appeared to be nearly the same at various places. It was usually better with farmers having better resources and in areas with better fodder production potential compared to landless labourers and in areas with low potential for fodder. Jersey breed was suited for upgrading of cattle of Central India farmers due to its small size and high fat content. It is also suited for comparatively scarce fodder conditions. Results of most of the studies on cross-bred animals obtained from army dairy farms indicated that the milk production efficiency of cross-bred cows declined with increase in exotic blood beyond 62.5 per cent. The National Commission on Agriculture had also recommended that in rural areas the aim should be to have crossbreds with 50 to 75 percent exotic blood. In some areas with poor fodder resources, crossbred cow possessing 25 per cent exotic blood may be more profitable (Gill and Handa 1981).

10.1.5. Cross-bred cow vs. buffalo

Reddy et al (1980) conducted a study on the comparative efficiency of cross-bred cow, non-descript cow and buffalo with regard to cost of milk production in and around Bangalore. The net daily cost of maintenance worked out to be Rs 6.43, Rs 2.82 and Rs 4.20 while the cost of producing a kg of milk was found to be 80.07, 105.57 and 109.38 paise, for a crossbred, a non-descript cow and a buffalo, respectively. Thus, the maintenance of cross-bred cows was found to be more economical due to better performance and high producing capacity compared to buffaloes and non-descript cows. Similar trends have been reported by Reddy and Mathur (1978) in a study on comparative performance of cross-bred cows and buffaloes at Agra, U.P. The utility of cross-breds with Jersey semen has also been established on farmers' fields in Indo-UK Dry farming Project areas (Indo-UK DFP 1980b).

10.2. OTHER LIVESTOCK BASED INTEGRATED FARMING SYSTEMS

As already stated, raising livestock was more remunerative than arable farming in arid and semi-arid regions particularly for marginal and small farmers or for those having slopy land with shallow soil, not suitable for raising crops. While most farmers in semi-arid and sub-humid regions have dairying as the subsidiary profession, rearing of sheep and goats was more common in arid regions due to their voracious eating habits and ability to travel long distances in search of food and water. Both sheep and goat can be reared together as their grazing habits are not competitive but complementary. While goats browse, sheep mainly graze. The natural vegetative cover on marginal, slopy, eroded and wasteland or appropriate vegetative cover to be created as soil conservation measure on such poor land, was sufficient to maintain a herd of sheep and goats. The common belief that sheep and goats were responsible for desertification was not supported by scientific data.

10.2.1. Goat farming

Goat was poor man's cow, although it was raised for both milk and meat. Goat survived on meagre quantities of feed, fodder and water. While Marwari breed of goat proved best for arid conditions, Jamunapari was best for semi-arid conditions. Most of the feed requirement of goats was met through grazing. Atriplex species of trees were valuable source of adequate quantity and quality of feed in arid and semi-arid regions (Kinet et al 1998). Other fodder trees, shrubs and grasses were also valuable source of feed to goat. Goats should be fed with concentrates like maize, horse gram, groundnut cake, wheat bran and fish meal. Common salt should also be added. Fresh water must be supplied as it helps digestion, blood circulation and removal of toxins from body. Water should be changed every morning and evening. Water also regulated body temperature. Goats should be provided dry, comfortable, safe and secure housing free from worms. The ideal space requirement was 4 to $5m^2$/goat. The effect of winter was more pronounced on the lactation period of goats as compared to summer months. The data on 90 days lactation yield, total lactation yield, lactation length, fat and SNF (solids not fat) content of milk revealed that lower milk yield with low fat contents was observed in goats housed without shelter, while the higher lactation milk yield with higher fat content was observed in goats housed in pucca structures (Kaushish 2006). Goats mature in about one year when both buck and doe are ready for breeding. Gestation period was about 150 days. Goat gives birth to 1-3 kids each time at intervals of about 245 days. The kids can be weaned after 30-40 days. The doe can be allowed in mating 45-60 days after kidding. Goat farming based integrated system was introduced and popularised for livelihood security in disadvantaged districts of Bundelkhand Region by Central Institute of Research on Goats (I.C.A.R.), Makhdoom, Mathura, under National Agricultural Innovations Project (NAIP 2012)

10.2.2. Sheep farming

Sheep were better adapted to diverse conditions than goats-temperate Himalayan region to dry western region to southern region. They consumed large quantities of roughage material and changed it into valuable products viz. lamb meat, wool, etc. Housing also need not be so elaborate as that for goats. However, it has to be protected from predatory animals and fencing should be at least 2 metres high. An adult sheep requires about 1.5 kg of leguminous hay per day for average body weight. Additional protein was supplied through concentrates such as groundnut cake, sesame cake, sunflower/safflower cake, etc. particularly when pastures are poor in legumes or are deteriorated due to drought. To meet the deficiency of minerals in diet, mixture of salt, powdered limestone and bonemeal should be fed. Unlike other farm animals, ewes in general did not come in heat at regular intervals throughout the year but were seasonal in this respect. Also, mostly ewes remained in heat for 21-39 hours. The optimum time of service was towards the end of heat period. The gestation period varied from 142 to 152 days. Sheep grew fully well at the age two years and were ready for breeding.

10.2.3. Poultry farming

Poultry was one of the fastest growing industry in the world, as consumption of poultry meat and eggs was increasing at about 5 percent rate annually. Till 1950, it was considered a backyard profession in India but it had become an important agricultural industry now producing eggs, broiler and chicken, the consumption of which was increasing continuously. Backyard poultry farming with cropping ensured sustainable livelihood not only to small and marginal farmers but even to landless labourers. For farmers it was very good source of subsidiary income, while for the landless it was very remunerative profession.

10.2.3.1. Breeds: Dual purpose hens had been bred for both eggs and meat to further the cause of rural and tribal poultry keepers. Their details are as follows.

Krishna J.: It was replica of indigenous hen with better economic viability. The bird had high tolerance against diseases and heat.

Dual Purpose Coloured Bird: This was a multi-coloured bird, quite suitable for backyard, semi-intensive and rural poultry keeping. It adopted well to adverse climatic conditions.

Vindhyachal Broiler: It was produced by adopting an alternate breeding strategy i.e. mating the normal bodied sire with dwarf dam. It was suitable for commercial broiler farming. It had high tolerance to diseases and heat and thus quite suitable for tropical climate.

Kadaknath: The bird was habituated to Jhabua hills region (Dhar and Jhabua districts) and was famous for delicious meat. It was also known as kalamasi meaning black meat. Its plumage colour was black, bluish black and golden. It was well adapted to adverse climatic conditions and had high tolerance to diseases. Kadaknath bird does not sit on its eggs for natural hatching. Therefore, the eggs of Kadaknath were mixed with eggs of deshi bird and the deshi bird was made to sit on these eggs. Thus, this was the method adopted for hatching the Kadaknath eggs. Kadaknath production provided a sustainable livelihood to not only marginal farmers but also to landless in Jhabua, Betul and Dhar districts of Madhya Pradesh (NAIP 2012).

10.2.3.2. Chief characteristics: The chief characteristics of above mentioned four breeds were presented in Table 61.

Table 61. Chief characteristics of various breeds

Characteristics	Breeds			
	Krishna-J	Dual Purpose coloured bird	Vindhyanchal broiler	Kadaknath
Chick weight	33 g	40g	40 g	28-34 g
Weight at 6 weeks	-	1 kg	1.5 kg	-
Adult body weight	1.3kg	1.8 to 2.0 kg	-	1.6 to 2.2 kg
Dressed weight (per cent)	-	78	81.8	-
Egg production	120 eggs/year	170 eggs/year	-	106 eggs/year
Average weight /egg	51g	56 g	-	42 to 50 g
Feed consumption	85 g/day	120 g/day	-	-

10.2.3.3. Feed: The feed conversion efficiency of poultry birds was superior to other animals. About 60-70 percent of the total expenditure on poultry farming was spent on the poultry feed. The feed should be balanced containing carbohydrates, fats, proteins and minerals. The common feed-stuffs used for making the poultry feed are: cereals, oil cakes, salts and minerals. From one-day old to 4 weeks of age, birds were fed on starter ration and there after finisher ration, which contained more energy and 18-20 percent protein. The feed was served 2-3 times a day. In addition to feed, antibiotics and some medicines (Riboflavin) were also added to poultry diet. Laying hens must be provided with oyster shell or powdered limestone.

10.2.3.4. Housing: Adequate space (0.3-0.4m^2 per bird) was required and the house should be well-ventilated with temperature control.

10.2.3.5. Management: Chicks must be vaccinated against Ranikhet disease at 6-7 days' age. When they are about 6 weeks old, they can be sold as broiler. Hens could be retained for one and a half years for production of eggs and then sold for table purpose. One hen could lay 180-230 eggs in a year starting from the sixth month. It also produces about 230 grams of fresh droppings per day.

10.2.4. Piggery

Pigs were kept for the production of pork. The pigs grew fast and were prolific breeder. The swine farrowed 10-12 pigs at a time and produced two litters in one year, if managed well. The carcass return was high - quantity of pork available was 65 to 70 percent of live weight. Thus, piggery was economically very sound component of a farming system, particularly for financially weak persons. Piggery required small investment and gave quick returns.

Pigs feed on forages, certain grain by-products from mills, meat and fish by-products, damaged feeds not fit for human consumption, etc. Locally available materials can be used to provide comfortable shelter to pigs, one big size pig requires about 3m^2 space and walls only 1.2m high. As pigs were the most rapidly growing livestock, feed management was very important for pig production. The complete diet of pigs consisted of stuffs containing protein, carbohydrates, fats, minerals, vitamins and ample good water. Pigs have a simpler stomach than ruminants and therefore they must be fed with maximum amount of concentrates and minimum amounts of roughage. The main ingredients of the feed should, therefore, be cereals, millets and their by-products, and 0.5 percent of added salt, calcium (Ca) and phosphorus (P). If agricultural/kitchen waste or fish and slaughter house waste are included in diet, addition of Ca and P-carriers are not required. The quantity of feed to be given should be 2.5 to 3.0 kg per 100 kg body weight.

Sows weighing about 100 kg at 12-14 months age were ready for breeding. Sows with low body weight showed higher mortality of piglets before weaning. The gestation period was on an average 144 days. The litter size at birth could be 1 to 16 in number. The time taken in the whole process of farrowing ranged from 2 to 4 hours. Sows are weaned after 40 days. Weaned sows came in to heat in 3-10 days after weaning. The boar-sow ratio should be 1:15 and it was profitable to raise two litters from each sow every year.

For successful piggery, care must be taken to avoid mortality of piglets due to disease/parasitic infection. Mortality in newly born piglets was maximum when farrowing took place in acute cold or hot weather. Therefore, breeding should be so planned that farrowing occurred during moderate season.

Piggery as a source of livelihood proved a grand success in Dumka and Jamtara districts of Jharkhand and in north eastern India (NAIP 2012).

10.3. BEE KEEPING

Bee keeping was an ideal activity for development as a subsidiary occupation providing supplementary income to rural, hilly and tribal people, particularly to farmers and horticulturists, because of rich flora available in rural areas. Bee keeping not only offers honey and other products but also enriches profusely flowering crops and horticultural plants in quality and quantity by crop pollination services rendered by bees. Bee keeping as an industry was of recent origin in India and received an impetus after the all India Khadi and Village Industries Board took its development on modern lines since 1953-54. Later on agricultural scientists got involved and consequently honey collection was now a modern industry based on scientific methods. The industry does not need any raw material as nectar and pollens from flowers are available in plenty as free gift of nature.

10.3.1. Species

Four species of honeybees are mainly found in India.

a. ***Apis dorsata*:** This was wild honey-bee about 20mm in size. This was the biggest species and produced more honey. It was found in almost all regions of India. It was very aggressive and dangerous. Its hives were found in high building, hilly regions, caves, trees, water tanks, etc. It can not be tamed and hence not suitable for bee-keeping.

b. ***Apis florea*:** It was smaller in size as compared to Indian bees, and found everywhere. Its hives were found in old houses, bushes and hollow tree-trunks and were smaller in size. It did not like darkness and therefore settled in open space. Its hives yielded 200 to 250 grams of honey each but it was sweeter. It was sensitive to cold and therefore it was not found at higher elevations-hills etc.

c. ***Apis cerana indica*:** It was entirely different from *Apis dorsata* and *Apis florea*. It was native to India, cool natured and about 15mm in size. It liked darkness and was very active. Its hives were also found in bushes, hollow stems, and ceilings of houses. Bee-keepers keep them in wooden boxes made for the purpose. It left its hive only in odd (unfavourable) circumstances.

d. ***Apis mellifera*:** This species was imported from Europe (Italy) being similar to *Apis cerana indica* in nature and size but yielded 9 to 10 times honey. Initially it was kept in northern India but was acclimatized to whole India later on. It was easily kept in wooden boxes and the quality of its honey was good.

10.3.2. How to start

Before starting, the farmer should get a training for it. Generally it can be started any time except extreme weather period. The bee-keeper should start with at least 2 and not more than 5 colonies. A minimum of 2 colonies was recommended so that the job was not lost completely if the queen in one colony died due to some mishap.

10.3.3. Hive

The hive consists of floor board, brood chamber, super chamber, top cover, inner cover, frames and entrance rod. These parts can be easily separated. The hive can be single or double walled, but the single walled hive is light and cheap.

10.3.4. Honey Collection

The most suitable time for commencing bee-keeping was early summer (February-March) or late spring (October-November). Honey quality varied in the proportions of nectars produced by different plants. The nectar collected by bees was processed and placed in comb cell for ripening. During ripening, sucrose was converted into glucose and fructose by the enzyme invertase added by bees.

10.3.5. Economics: (Tomar and Verma 2014)

Annual cost of 5 colonies	: Rs. 23,500
Estimated Income	: Rs. 45,000
Benefit: Cost ratio	: 1.9

10.4. LAC CULTIVATION

Lac was natural substance secreted by a micro female insect to form an external coat on its body. The secretion comes out from lac glands of the female lac micro-insect in liquid form but on drying, the same turns into a hard solid state. The female insect gives out lac for the protection of herself and the offsprings. The lac glands open into upper surface of the body as many pores. Besides these, 3 wax glands are situated on mouth, a respiratory pore in central part of its chest and on edges of its anus. Wax was secreted continuously from these glands and the same did not allow lac to be deposited on respiratory pores and on anus and kept the lac micro-insect alive. The lac insects lived on the juice extracted from the branches of host plant. The colour found in the blood of lac insects was also important for the industry. Lac cultivation has emerged recently as a very remunerative subsidiary occupation for farmers. The technology for lac cultivation has been evolved by Central Lac Research Institute (ICAR), Ranchi. Lac cultivation has been extended to farmers' fields under National Agricultural Innovation Project, Component III, in two disadvantaged districts of Madhya Pradesh viz. Betul and Mandla. Lac cultivation was previously limited to palas trees (*Butea monosperma*) but now it has been tried successfully on ber (*Ziziphus jujube*) trees also (Sahu et al 2014b). These trees can be easily planted on field bunds and used for lac cultivation as gainful employment particularly to small and marginal farmers.

10.4.1. Varieties

There were two varieties of lac viz. Kusami and Rangini, and two crops a year of both these were raised. The crops of Rangini lac were raised in October-November and in April-May where as those of Kusami in December-January and June-July. There are two methods of lac cultivation: traditional method and scientific method. The traditional method was very old and unscientific, as the lac was obtained by cutting the host plant (palas) itself, and no attention was paid to lac inset. Consequently, the lac insects died and future crop of lac was damaged. Presently this method was adopted by untrained farmers only. Modern method was based on research work conducted by Indian Lac Research Institute, Ranchi (Jharkhand). In this method, bringing up the lac insect on host plant for lac cultivation did not harm at all the host tree. The method consisted of three steps: (i) pruning of host plant, (ii) inoculation of host plant with brood (lac seed) carried on sticks or in bags, and (iii) harvest of lac crop. For starting lac production, the host tree (palash, ber, kusumi and rain tree) has to be prepared/shaped. The tree was pruned lightly, so that new soft and more juicy branches were produced. The pruning of palash (*Butea frondosa*) was done in the month of March-April, and that of ber and kusumi in December. Pruning was done so that all dry and diseased branches were removed. Also, for pruning a sharp and good implement must be used so that the cut was slanting, clear and no cracking in branch-wood was observed. Three/four pieces of a brood lac stick (seed) about 15cms long were bundled together. While making bundle, both the ends of rope should be free (about 10cm long) so that these could be used to tie the pieces of brood sticks to the host tree. This process was actually the inoculation of host tree with lac seed (brood), which consisted of micro insects coming from mature female lac insects.

10.4.2. Care of Lac Crop

After inoculation of host plant, the lac insects had to be protected from frost/fog damage and insect pests. To control these, spray of pesticide/fungicide was done, if necessary and unavoidable to save the crop. For palash lac, Ethophanprox (0.02 per cent) or Phipronil (0.07 percent) or Indoxicarb (0.05 percent) with Carbandizim powder (0.05 percent solution) spray was recommended.

10.4.3. Harvesting

Harvesting of lac crop could be done any time after maturity but for brood, harvesting in the month of October had to be done when lac insects start coming out. Also, only those branches which have a continuous layer of lac should be harvested. Branches with sparse deposit of lac should be left out so that brood spreads to vacant space and April harvest produced higher yield of lac.

10.4.4. Economics

Usually palash tree was used for lac cultivation and two crops per year-one in summer and the other in winter were raised. A total of 770 palash trees were planted on one ha land maintaining a diagonal distance of 3.5m x 3.5m. Thus, lac cultivation could be undertaken on 385 trees per season. The cost of cultivation/ha was Rs. 90,000 per ha- Rs. 40,000 on labour and Rs. 50,000 on other items. The produce obtained was 2400kg brood and 300kg lac per ha. The estimated value of the produce was Rs. 2,85,000. Thus, a net profit of Rs. 1,95,000 per ha was obtained (Sahu *et al* 2014a).

10.4.5. Adoption

The lac cultivation has been popularised in several states under National Agricultural Innovation Project, component III, by conducting training programmes and demonstrations on farmers fields. Jawaharlal Nehru Krishi Vishavidyalaya, Jabalpur, introduced lac cultivation on palash and ber host plants in Betul and Mandla districts extensively. Almost every farmer in Betul district obtained 60kg lac and a net profit of Rs. 3500 by lac cultivation on 20-25 trees of palash/ber owned by them. Similarly, Krishi Vigyan Kendra, Basantpur, District Rajnandgaon, Chhattisgarh, extended lac cultivation in Rajnandgaon district and the estimated average income was about Rs. 2 lakhs per ha. Also, in Jamtara district of Jharkhand state, livelihood security through lac cultivation was assured for small and marginal farmers (NAIP 2012).

10.5. LIVELIHOOD SECURITY THROUGH MUSHROOM PRODUCTION

Production of mushroom is a profitable job to assure livelihood security. Mushroom is a type of fungus of which some varieties are edible. Oyster mushroom is one such very popular and easily cultivable variety. Its production is a boon to landless or marginal farmers and for farm women's empowerment. It can be easily undertaken in rural areas using agricultural waste materials such as paddy and wheat straw. For a small unit, a space of 6m x 5m x 3m was adequate.

10.5.1. Method of production

For production of Oyster mushroom, an enclosure made of bamboos and polythene, or a hut with mud-walls and thatched roof was required. The structure should have cross ventilation through windows and wooden doors with iron wire-net. The structure of size 6m x 5m x 3m can accommodate 300 bags. At first, the cellulosic material (paddy/wheat straw) has to be sterilised because it could be contaminated by bacteria, undesirable fungi, and other micro-organisms, even though fresh. Before sterilisation, the material should be cut into pieces of 2 to 3cms size. For sterilisation treatment with hot water (60-65°C) or chemicals like bevestin (5gms/10litres) or formalin (125ml/10litres) was done. When the moisture content of this sterilized media was 60 to 70 percent, mushroom spores were mixed into it. After this, about 5kg of the material mixed with spores, was put into plastic bag each. Every bag should have holes all over about 5mm in diameter to assure aeration and control of temperature. The polythene bags were stored at 25°C for 15 to 20 days. First crop was ready for harvest in 6-7 days when the top of mushroom was 8 to 10 cms in diameter. Two more crops from each bag were harvested subsequently at intervals of 5-6 days.

10.5.2. Economics of production

The yield of mushroom obtained was 700 to 800 grams per kg of dry straw used, i.e. about 0.4kg per polythene bag or 120kg per unit of production. The approximate investment per unit was Rs. 4,000 and gross income Rs. 9,000/. Thus, the net income per unit came to Rs. 5,000/- and the benefit: cost ratio was 2.25.

10.5.3. Extension of Technology

The technology was extended to villagers under National Agricultural Innovation Project (ICAR) 2006-13 as livelihood security measure with successful results as follows.

a. Mushroom cultivation was at first introduced in Balaram village and then extended to whole Dhalai district of Tripura state.
b. Mushroom production proved to be a very profitable means of livelihood in Samastipur district of Bihar.
c. Production of Oyster mushroom on paddy straw was a grand success in Bhubaneshwar district of Odisa.
d. Mushroom production was introduced in Mudiyani village of Champawat district (Uttarakhand) and became very popular with small and marginal farmers (NAIP 2012).

10.6. FARM POND/TANK BASED INTEGRATED FARMING SYSTEMS

As already pointed out, collection of run-off water into farm pond/tanks and its recycling could completely change the status of dry land farmers in semiarid and sub-humid regions. It could save rainy season crops from drought, could make double cropping a success and provide livelihood security and prosperity to marginal farmers/landless labourers through farm pond/tank based integrated farming systems. Some of these have already been discussed in chapter VI, and are summarised in continuation.

10.6.1. Rice-Fish-Poultry System

The system has met grand success in Tamilnadu (NAIP 2012) but holds promise over large areas under sub-humid and humid regions of Central India.

10.6.2. Rice-Fish-Vegetable Systems

The system proved excellent in Assam (NAIP 2012) but holds promise in humid and sub-humid regions, through out the country, particularly in eastern U.P., Bihar and Chhattisgarh.

10.6.3. Rice-Fish-Piggery/duck farming System

The system had met a great success in Assam and Meghalaya, and was very promising for rice growing regions of Central India.

10.6.4. Rice-Fish-Makhana (*Euryale ferox* Salisb) and water-chestnut (*Trapa bispinosa* Natans.) System

It was strongly recommended for flood prone ecosystem of Darbhanga district of Bihar.

10.7. INTEGRATED FARMING SYSTEMS FOR LIVELIHOOD SECURITY IN RAINFED MICRO-WATERSHEDS OF JHARKHAND

Depending upon available land, human power in the family and the attitude of the people, four models were recommended for Dumka and Jamtara districts of Jharkhand state. The income earned from these models, over and above the traditional system adopted by them, was presented in Table 62.

Table 62. Income from Different Integrated Farming Systems (NAIP 2012)

S. No.	Farming System	Available Agricultural Land (ha)	Additional Income Rs.
1.	Poultry + Piggery + Goatry	Landless	17579.0
2.	Poultry + Cropping + Goatry	Upto 1.0 ha (marginal farmers)	19668.0
3.	Poultry + Cropping +Horticulture + Lac +Aquaculture + Goatry	1.0 to 2.0 ha (small farmers)	24155.0
4.	Poultry + Cropping+ Horticulture + Lac +Goatry+ Aquaculture + Piggery + Duck	2.0 to 4.0 ha (medium farmers)	39565.0

■■■■

BIBILIOGRAPHY

AICRPDA 1980 *Salient research results of Dryland Research Project sub-centre Rewa* (1971-79) Jawaharlal Nehru Krishi Vishwa Vidyalaya campus College of Agriculture, Rewa, Madhya Pradesh.

AICRPDA 1983 *Improved Agronomic Practices for Dryland Crops in India,* 3rd Edition. All India Coordinated Research Project on Dryland Agriculture (ICAR), Hyderabad AP (now Telangana).

AICRPDA 1994 *Progress Report, Integrated research project for soil conservation and watershed management,* All India Coordinated Research Project on Dryland Agriculture, Jawaharlal Nehru Krishi Vishwa Vidyalaya (now Rajmata Vijayaraje Scindia Krishi Vishwa Vidyalay) campus College of Agriculture, Indore, Madhya Pradesh.

Alam A 1994 Design, development and adoption of farm equipment (Un-published paper).

Arnon I 1975 Physiological Principles of Dryland Crop Production in *Physiological Aspects of Dryland Farming* (ed) U S Gupta, Oxford & IBH, New Delhi, p. 3-124.

Ayres Q C 1936 *Soil Erosion and its Control.* Mc Graw Hill Co. New York, U.S.A.

Bali J S 1974 *Soil degradation and conservation problems in India.* UN-FAO Consultation of Experts on Soil Degradation, Food and Agriculture Organization, Rome, Italy.

Bansil P C 2001 *Agricultural Statistics in India,* 4th Edition, CBS Publishers and Distributors, New Delhi, pp 246-247.

Berry Wendell 1981 The making of a marginal farmer. *Span,* June 1981 pp 5-9.

Bhargava O D, Russell M B and Verma G P 1973 Run-off from maize and soybean plots and potential of double cropping under un-irrigated conditions at Jabalpur. *Indian Journal Agricultural Science* **43** 129-132.

Bhargava O D, Verma G P, Russell M B and Katre R K 1976 Soil moisture depletion pattern under maize and soybean crops grown on deep black soil. *Journal of the Indian Society of Soil Science* **24** 158-162.

Biswas P P and Sharma P D 2005 Rainfed Agriculture- some strategic interventions. *Indian Journal of Fertilizers* **1** 25-34.

Blevins R L 1986 Reduced tillage and soil properties, in *Proceedings International Symposium on Minimum Tillage.* Bangladesh Agricultural Research Council, Dhaka, Bangladesh. pp 34-44.

CGWB 2006 Hydrology Project. Central Ground Water Board. Ministry of Water Resources, Government of India. New Delhi.

Chandra S K 2006 Efficient management of land resources for sustainable agriculture in Bihar. *Journal of the Indian Society of Soil Science* **54** 435-442.

Chaturvedi M C 1985 Water resources of India-an overview. *National Seminar on Water Resources, Systems and Planning.* Indian Academy of Sciences, Bangalore, Karnataka.

Chowdhury S L 1979 Fertilizer management in dryland regions for increased efficiency. *Fertilizer News* **24** 61-66, 101.

CRIDA 1987 *Research Highlights.* Central Research Institute for Dryland Agriculture, Hyderabad A. P. (now Telangana).

CRIDA and IASRI 2012 *Prioritization of Rainfed Areas.* Study Report 4, Central Research Institute for Dryland Agriculture, Hyderabad and Indian Agricultural Statistics Research Institute, New Delhi. National Rainfed Area Authority, Planning Commission, Government of India New Delhi, February 2012.

CSARD 1998 Watershed Management Technology for Reclaiming Ravineous Land. Final Report, Department of Science and Technology, Government of India Project (1994-98) Centre of Sustainable Agriculture and Rural Development, Gwalior, M. P.

CSWCRTI 1980 *Project for drought, flood and sediment control- Sukhomajri.* Central Soil and Water Conservation Research and Training Institute, Research Centre (ICAR) Chandigarh, Haryana.

Dastane N G 1968 New concepts, practices and techniques in the field of water use and management. *Symposium on Soil and Water Management.* Indian Council of Agricultural Research, New Delhi.

Department of Agriculture 2009 *Agricultural Statistics of Chhattisgarh State 2009.* Department of Agriculture, Government of Chhattisgarh, Raipur Chhattisgarh.

DES 1979 *A manual of available technology for raising farm production.* Directorate of Extension, Jawaharlal Nehru Krishi Vishwa Vidyalaya, Jabalpur, Madhya Pradesh.

Deshpande W R 1973 *A Handbook of Agriculture.* Directorate of Extension. Jawaharlal Nehru Krishi Vishwa Vidyalaya, Jabalpur, Madhya Pradesh.

DFW & AD 2009 *Agriculture Statistics.* Department of Farmers Welfare and Agriculture Development, Government of Madhya Pradesh, Bhopal, Madhya Pradesh.

Dubey U C and Singh K P 2014 Machinery for Improved Crop Production. *Krishi Vijay 2014-Farmers Fair Special,* All India Farmers Fair and Agricultural Exhibition. (27-29 September, 2014) Directorate of Extension, Rajmata Vijayaraje Scindia Krishi Vishwa Vidyalaya Gwalior, Madhya Pradesh.

Gadkary D A 1966 *A manual on soil conservation.* Department of Agriculture, Government of Maharashtra, Pune, Maharashtra.

Gajri P R, Majumdar S P and Sharma P K 2012 Tillage. in *Fundamentals of Soil Science* (eds) N N Goswami, R K Rattan, G. Dev, G. Narayanasamy, D K Das, S K Sanyal, D K Pal and D L N Rao, New Delhi; Indian Society of Soil Science, pp 161-179.

Gill G S and Handa M C 1981 Strategy for Dairy Cattle Breeding Improvement in India. *Livestock Advisor* **6** 3-4.

Gupta J P 1980 Effect of mulches on moisture and thermal regimes of soil and yield of pearlmillet. *Annals of Arid Zone* **19** 132-138.

Gupta R K 2002 Natural resources conservation technologies for black clay soil region. *Journal of the Indian Society of Soil Science* **50** 438-446.

Gupta Ram K and Verma S K 1983 Water behavior of black clay soil as influenced by degree of sodicity. *Current Agriculture* **7** 114-121.

Gupta R K, Sharma S G, Tembe G P and Tomar S S 1978 A new approach in farming system for problem soils of Central Madhya Pradesh. *JNKVV Research Journal* **12** 73-79.

Gupta R K, Tomar S S and Tomar A S 1979 Improved soil management practices for maize and sorghum grown on Vertisols of Central India. *Journal of Agronomy and Crop Science* **148** 479-483.

Halvey A H and Kessler S H 1963 Increased tolerance of bean plants to soil drought by means of growth retarding substances. *Nature* **197** 310.

Harrington L W, Fujisaka S, Morris M L, Hobbs P R, Sharma H C, Singh R P, Chaudhary M K and Dhiman S D 1993 Wheat and Rice in Karnal and Kurukshetra Districts, Haryana, India. In Farmer's Practices, problems and Agenda for Action. CCS Haryana Agriculture University, ICAR, CIMMYT and IRRI Publication.

Hazra C R 1995 *Advances in Forage Production Technology*, Indian Grassland and Fodder Research Institute, Jhansi (Uttar Pradesh).

Hudson N W 1988 Conservation practices and run-off water disposal on steep lands, in *Conservation Farming on Steep Lands.* (eds) W C Moldenhauer and N W Hudson, World Association of Soil and Water Conservation, Ankeny, Iowa. pp 117-128

Hurni H 1986 *Guideline for development agents on Soil Conservation in Ethiopia.* Community Forests and Soil Conservation Development Department. Ministry of Agriculture. Addis Abeba, Ethiopia. pp 1-100.

Hurni H 1988 Options for Conservation of Steep Lands in subsistence Agricultural systems, in *Conservation Farming on Steep lands.* (eds) WC Moldenhauer and NW Hundson. Soil and Water Conservation Society, Ankeny, Iowa, U.S.A. pp. 33-44.

I A R I 1976 *A New Technology for Dry Farming.* Indian Agricultural Research Institute. New Delhi, p 121.

I A S R I and I C A R 2005 *Agricultural Research Data Book.* Indian Agricultural Statistics Research Institute and Indian Council of Agricultural Research, New Delhi.

ICAR 1976 *Crop Life Saving Research*, ICAR- IDRC Seminar Proceedings Indian Council of Agricultural Research, New Delhi.

ICAR 1977 *Dryland Agriculture Research Progress Report* (1970-75) Indian Council of Agricultural Research, New Delhi.

ICAR 1987 Vetiver grass (*Vetiveria zizanoides*) – A method of vegetative soil and moisture conservation. ICAR – US Advertising Private Ltd., Jangpura Extension, New Delhi.

ICRISAT 1974 *Proceedings International Workshop on Farming Systems.* 18-21 November 1974 International Crop Research Institute for Semi-arid Tropics, Hyderabad, Andhra Pradesh (now Telangana) India.

Indo-UK DFP 1979 *Brief Progress Report (1974-79) Indo-UK Dry Farming Project* (ICAR) September 1979 Jawaharlal Nehru Krishi Vishwa Vidyalaya (now Rajmata Vijayaraje Scindia Krishi Vishwa Vidyalaya) Campus College of Agriculture, Indore, Madhya Pradesh.

Indo-UK DFP 1980a Gully control. Indo-UK Dry Farming Project (ICAR) Jawaharlal Nehru Krishi Vishwa Vidyalaya (now Rajmata Vijayaraje Scindia Krishi Vishwa Vidyalaya) Campus, College of Agriculture, Indore, Madhya Pradesh.

Indo-UK DFP 1980b Final Report (Phase-I) Indo-UK Dry Farming Project (ICAR) 1974-75 to 1979-80. Jawaharlal Nehru Krishi Vishwa Vidyalaya, Campus (now Rajmata Vijayaraje Scindia Krishi Vishwa Vidyalaya) College of Agriculture, Indore, Madhya Pradesh.

Irrigation Commission 1972 *Ground water potential-a review.* Ministry of Water Resources, Government of India, New Delhi.

Jain LK, Nema RS and Agrawal SK 1998 Field evaluation of few selected seed-cum-fertilizer drills in dryland Vertisols. *National Seminar on Technological advances in Agriculture and Livestock production in Madhya Pradesh.* 20-22 April, 1998 Jawaharlal Nehru Krishi Vishwa Vidyalaya, Jabalpur, Madhya Pradesh.

Jain M P 2012 Conservation agriculture in rainfed system, in *Conservation Agriculture* 2012 (eds) S S Tomar, Shobhana Gupta, S S Bhadauria, K V Singh and Neeraj Hada, Directorate of Extension, Rajmata Vijayaraje Scindia Krishi Vishwa Vidyalaya, Gwalior, Madhya Pradesh.pp 56-93.

Jain M P, Sharma S K, Thakur H S, Swarup Indu and Agrawal S K 2014. *Forty Years (1971-2010) of Dryland Agriculture Research and Technology Development at Indore Centre.* All India Coordinated Research Project on Dryland Agriculture, Rajmata Vijayaraje Scindia Krishi Vishwa Vidyalaya Campus College of Agriculture, Indore, Madhya Pradesh.

Kale V S 1977 Occasional Paper on Run-off Management (unpublished).

Kampen J, Tejwani K G, Singh S N, Prihar S S and Spratt E D 1976 *Glossary of Technical Terms for Land-use and Crop Production in Rainfed Areas.* Indian Council of Agricultural Research and